The Stanleys of Newton

Yankee Tinkerers in the Gilded Age

by Karen H. Dacey

Stanley Museum, Inc., Publisher
Kingfield, Maine

The Stanleys of Newton: Yankee tinkerers in the gilded age
by Karen H. Dacey

For further information, contact:
The Stanley Museum, PO Box 77, Kingfield, Maine 04947
Telephone: 207-265-2729, Fax: 207-265-4700
Email: maine@stanleymuseum.org, Web site: www.stanleymuseum.org

A branch of the Stanley Museum of Maine:
The Stanley Museum of Colorado
517 Big Thompson, PO Box 788, Estes Park, Colorado 80517
Telephone: 970-577-1903, Fax: 970-577-1924, Email: estespark@stanleymuseum.org

First edition: October 2009

Library of Congress Cataloging-in-Publication Data

Dacey, Karen H.
  The Stanleys of Newton : Yankee tinkerers in the gilded age / by Karen H. Dacey.
     p. cm.
  Includes bibliographical references and index.
  Summary: "A biography of Francis E. and Freelan O. Stanley, identical twins
and inventors of the Stanley steam car, and the history of their lives in Newton,
Massachusetts, from 1890 to 1940"--Provided by publisher.
  ISBN 978-1-886727-15-1 (pbk. : alk. paper) -- ISBN 978-1-886727-16-8
(hardcover : alk. paper)
  1. Stanley, Francis Edgar, 1849-1918. 2. Stanley, Freelan Oscar, 1849-1940.
3. Automobile engineers--United States--Biography. 4. Stanley Steamer automobile. I.
Title.
  TL139.D33 2009
  629.222092'273--dc22
  [B]
                              2009037632

Typeface: body copy Adobe Garamond, chapter headings ITC Benguiat
Cover and Original artwork © 2009 by Stanley Museum, Inc.

Book designed by Deb McGee of D.J. McGee Graphic Design, Presque Isle, Me.
Cover design by Jen Libby of Carrabassett Marketing & Printing, No. Anson, Me.

Dedication
For Tim

Table of Contents

LIST OF ILLUSTRATIONS

Chapter One

Chansonetta Stanley Emmons' circa 1898 photograph of the Stanley homestead in Kingfield suggests the cozy charm that captivated Mrs. F.E. Stanley.

A panoramic view of Kingfield, population about 600, taken circa 1898 by Chansonetta Stanley Emmons.

The Stanley Twins, Inc., around nine years old – the age when they launched their first mechanical enterprises.

The Stanley twins' signatures as fixed on a 1901 joint patent application.

Apphia Kezar French.

Frank Stanley and his "Gustie" at the time of their marriage in 1870.

In this crayon portrait of his four-year-old daughter Blanche, Frank Stanley used his atomizer and a piece of lace as a stencil to spray on the collar of her dress.

Photographic portrait of the artist in his Lisbon Street studio in Lewiston.

Chapter Two

Six years after his brother married, Freelan wed Flora Jane Record Tileston.

The Stanley Practical Drawing Set: a compass, ruler, protractor, right triangle and square, with carrying case.

A business bill head from F.O. Stanley's Mechanic Falls shop.

Show them and they will buy.

Chapter Three

Sterling Elliott, inventor and mechanical master, was F.O. Stanley's lifelong friend. Elliott manufactured bicycles in Watertown and later became a leading force in the "Good Roads Movement."

When they first moved to Massachusetts Freel and Flora rented this house, owned and built by Sterling Elliott, at 24 Maple Street in Watertown.

Before the tracks of the Boston and Worcester RR were depressed, man, locomotive, and beast traverse Newton Corner in parallel lines.

F.O. Stanley built his first Newton home at 165 Hunnewell Avenue in 1894 – quite a step up from Maple Street.

The F.E. Stanleys moved to their third and most elegant Newton address, 638 Centre Street, in 1896.

Newton High School in Raymond Stanley's day.

Chapter Six

George Eastman and Mrs. Pauline Abbott take a back seat to Kodak vice president Charles Abbott at the wheel of a Locosurrey.

Sylvester Roper of Roxbury, Massachusetts, built this steam-powered motorcycle in 1868 – just three years after the end of the Civil War.

View of P. A. Murray's carriage shop in Newton Corner looking west on Washington Street.

"Twin-powered" single engine: the Stanleys pose in their first steam car.

Chapter Seven

Mechanics Hall as it looked in 1906 on Huntington Avenue, Boston.

This Bert Poole illustration shows the bicycle track at Charles River Park, scene of F.E. Stanley's first steam triumph.

Chapter Nine

Without a windshield, without a top, and certainly without seatbelts, Flora and Freel made it unscathed to the top: "…(F)or a gentleman and his wife to ride up Mt. Washington by such a vehicle will cost less than 25 cents."

The Tally Ho Coach: eight-horse-power ride to the summit.

Chapter Eleven

The Stanley Motor Carriage Company at 44 Hunt Street.

F.E. Stanley and driver Fred Marriott pose with one of the Vanderbilt racers.

Following his outstanding (but not winning) run up Mt. Washington, Frank Stanley maneuvers his Model B runabout down wooden planks over the steps to the platform at the Summit Lodge.

Steam on steam: F.E. Stanley and Joe Crowell betray nothing of their eight-mile jolting ride to the top of Mt. Washington as they sit stoically for a picture with an engine from the Mt. Washington cog railway.

Artist Peter Helck captured the excitement of William Hilliard's' triumphant finish in the 1905 Climb to the Clouds race.

In 1905 F.E. Stanley again challenged the precipitous course, this time in a Model H speedster.

Chapter Twelve

Before the start of the mile trials competition Fred Marriott checks out the "wog."

Marriott takes a test run on the beach.

Chapter Thirteen

Chapter Fifteen

Chapter Sixteen

Raymond Stanley as a corporal in the Signal Corps during World War I.

The Stanley daughters, Blanche (left) and Emmie, in an 1892 photograph.

F.E. Stanley took this portrait of his sister Chansonetta in 1888.

Chapter Seventeen

Though smaller in scale, the Stanley Hotel in Estes Park recalls the grace and grandeur of the White Mountain resort hotels.

A view of the main lobby at the Stanley Hotel.

Chansonetta Emmons took this photograph of F.O. Stanley's grand last Newton residence built in 1913.

Gustie and Frank "roughed it" at their Squirrel Island cottage.

Here seen blending with the surroundings F.O. Stanley, while recuperating in the Rockies, carved out a life quite independent of his brother.

F.O. Stanley lived twenty-two years longer than his twin.

Newton Fire Chief W.B. Randlett sits next to his son, C.B. Randlett (at wheel) in a 1913 Model 78 Stanley with Fire Department insignia, extinguisher, and bell.

F.E. Stanley.

F.O. Stanley.

The Stanley Museum's
Mission Statement

The Stanley Museum keeps and shares the values of Yankee ingenuity and creativity as exemplified by the Stanley family in order to inspire these values in children and adults.

Institutional Vision & Values

The Stanley Museum shall be for all an institution of learning dedicated to a celebration of Yankee ingenuity as exemplified by the Stanley twins, Freelan Oscar and Francis Edgar, their sister Chansonetta Stanley Emmons, and their American contemporaries.

The Stanley Museum shall strive to preserve the history of their achievements and the artifacts and technology relevant to them for the purpose of arousing in the public pride of heritage and promoting those values most cherished and valuable in the American experience – Family Excellence, Integrity, Creativity, a Love of Learning, Tenacity, and Good Citizenship.

Adopted Unanimously October 14, 2000

Dr. Gerald Entine

The Stanley Museum is deeply grateful to Dr. Gerald Entine, who contributed the funds to publish *The Stanleys of Newton: Yankee tinkerers in the gilded age.*

Dr. Entine is the founder and president of Radiation Monitoring Devices, Inc. that now occupies the former Stanley Motor Carriage Company factory building.

Dr. Entine's creative work in developing new technologies and bringing them to market mirrors the work of the Stanley Twins and their sister, Chansonetta.

Dr. Entine holds a B.SC. in Physics/Biophysics and an M.A. in Physics from the University of Pennsylvania and completed his Ph.D. in physics from the University of California at Berkeley under the direction of two Nobel Laureates.

In 1971, he began development work on an infrared photosensor, which became the first practical device made from CdTe. In 1972, he turned his attention to a new, solid state nuclear detector made of CdTe, and assumed responsibility for all aspects of the work, including crystal growth, detector fabrication, and marketing activities. This technology provided the basis for Radiation Monitoring Devices, Inc., which Dr. Entine spun off from Mobil Tyco Solar Energy Corp. in August 1974, and has led to a wide variety of advanced nuclear instruments for medicine, power plants, and industry.

Since 1974, Dr. Entine has spent his time and effort guiding RMD as its founder and president. Dr. Entine continues to be involved in research, and has been the principal investigator on numerous research contracts and grants funded both privately and by federal agencies. These programs included the development of a portable cardiac ventricular function monitor with the Nuclear Medicine Department of Massachusetts General Hospital, research on new nuclear techniques to measure peripheral blood circulation in diabetics with State University of New York in Buffalo, improvements on cerebral blood flow measuring systems with solid state sensors at New York University and Bowman Gray School of Medicine, as well as programs with Yale, Massachusetts General Hospital Department of Cardiology, and the West Roxbury Veterans Hospital. These programs have resulted in technical papers and presentations, and several extensive clinical programs. Dr. Entine is currently an Adjunct Research Assistant Professor in the Department of Neurology at the Bowman Gray School of Medicine.

Thank you Dr. Entine!

Publisher's Foreword

At its inception in 1981, the Stanley Museum began a tradition of popular and scholarly publications designed to bring the Stanley Family and their American values to life. This is the mission of the Stanley Museum and our very reason for existence.

As the list of Stanley Museum publications at the end of this book document, the Stanley Museum staff and trustees have succeeded in carrying out the Museum's mission through this publications program. Few museums have a national award winning *Quarterly* newsletter of such exceptional quality. The Stanley Museum now communicates electronically with its monthly eNewsletter, the *Winker*. Our Chansonetta Stanley Emmons Strategic Plan has just brought her 1909 booklet, *The Old Table Chair*, back into print in its centennial year.

This is an enviable record of exceptional accomplishment, rarely matched by an organization with such limited resources.

So it is with the long-delayed publication of Karen Dacey's book that the Stanley Museum builds on and continues this tradition. This project began in 1997 and unhappily was delayed for various reasons. This book continues the Stanley Museum's pattern of partnering with outside humanities scholars and combining their perspective with the expertise of its staff, just as it did with Kit Foster's book, *The Stanley Steamer*, and James Pickering's *Mr. Stanley of Estes Park*.

Karen Dacey brings *The Stanleys of Newton* to life and provides a fresh interpretation of these influential and quintessentially American Yankees. This is an important book because it takes the Stanleys beyond the steam car, for which they are primarily known, and helps to interpret their lives and the era in which they lived.

Much of what the Stanley Family has to teach us still lies buried in the Raymond Walker Stanley archive and other repositories. Karen Dacey inspires the Stanley Museum staff and trustees to continue to research, write, and publish this history.

As you read Karen Dacey's well-researched book, you will be inspired not merely by the Stanley Family's accomplishments, but also by the way they engaged and contributed to their community.

H. James Merrick Donald R. Hoke, Ph.D.
Stanley Museum Archivist Stanley Museum Consulting Director

Author's Acknowledgments

Over a decade ago Susan Davis, then director of the Stanley Museum, launched an ambitious project to tell the story of the American steam car, in particular to chronicle the lives of its most famous inventors. This book was one of three commissioned for the task.

In no small way, I owe thanks to H. James Merrick, archivist of the Stanley Museum, who vetted the text with a meticulous eye and an encyclopedic knowledge of the subject matter. Jim is also responsible for shepherding this book into print.

I am also grateful to two Newton historians, the late Thelma Fleishman, author of *Newton* in the "Images of America" series, and Susan Abele, curator of documents and photographs at the Newton History Museum, for their cheerful willingness to read and emend the passages on Newton social history.

Thanks to Timothy Dacey, my husband, who lent his considerable editing skills to the sharpening of the text.

Finally, a posthumous nod to Augusta Stanley, redoubtable diarist and wife of Francis E. Stanley—one of the inventing twins: without her tireless logging of daily events, I doubt whether this book could have been written.

Karen Dacey

Introduction to an Era

THIS IS THE STORY of the twin brothers who built the famous steam car. They did not invent the steam automobile, but they did produce a model that was as appealing to the first generation of car owners at the turn of the twentieth century as Henry Ford's Model T was to a later motoring public. Even today "Stanley Steamer" is a household word, though no longer a household item. Had steam power outlasted internal combustion in the early race to dominate automotive technology, the Stanleys, Francis and Freelan, might be remembered today in the same league with the Wright Brothers, and their adult home in Newton, Massachusetts, celebrated as an automotive Kitty Hawk. The ascendancy of the gasoline engine consigned them instead to a side rail of history.

Yet the story of the Stanleys is worth more than a footnote. In many ways their lives are more representative of their class and times than those of the highly celebrated inventors of the Gilded Age. Born in rural Maine at the mid-nineteenth century, self-taught engineers, pioneers in two separate industries, the Stanley brothers numbered in the ranks of America's newly fledged middle class technocracy. They were part of the generation that left the farm for lives of comparative wealth and growing complexity. Drawn in by the gathering force of the Industrial Revolution, they in turn altered life for generations of Americans to come.

When the Stanleys were boys in the 1850s, most Americans still grew the food they ate and made the clothes they wore. Middle class in the United States largely meant, as it had in the previous century, farmer, craftsman, and merchant. Just about every white male considered himself middle class or about to become so. For without the entrenched hierarchies of Old World aristocracy, an American man could conceivably expect to rise in station (Hofstadter 1971, 132). To be sure some individuals attained greater wealth and there were pockets of elitism in various parts of the country, but social status was not based solely on how large a fortune one amassed. As the history of the Stanley family in Kingfield, Maine, attests, a man's prominence often depended upon how well he embodied the values of his community.

Yet even as the twins were growing up in rural simplicity, the Industrial Revolution had started to change how Americans fed and clothed themselves. By mid-century steam power made it possible to locate mills and factories in cities where owners could take advantage of cheap immigrant labor. The application of science and technology to production increasingly meant that machines could do more of the work and in greater volume. In a remarkably short time, changes in the production of goods reconfigured how people lived and consequently what they believed in. After the Civil War, a different sort of middle class—engineers, managers, businessmen, bankers—emerged to run the new economy. While some of them became wealthier than anyone had previously imagined, for the first time other Americans, many of them foreigners who had come expectantly to the United States, fell into the growing class of unskilled laborers with little hope of advancement.

When manufacturing moved from the millstream to the city, so too did droves of native-born young people. The population of American cities began to swell not only with immigrants but also with rural youth who might never have left the farm just a generation before. In the post-Civil War era, more of the American countryside was under cultivation, but technological advances meant fewer hands were needed to do the work. The mechanization of agriculture, as well as mass production of goods, increased the flow of young people from the farm to the city. In the long-settled region of New England where the twins grew up, available farmland was by this time extremely scarce, and younger siblings often left the country because they could not find land to work (Schlereth 1991, 36-37). The fact that the Stanleys had an older brother to take over the family homestead in Maine naturally predisposed the twins to test the wider world.

Between 1870 and 1900 the urban labor force in America grew to comprise more than one third of the population, creating a glut of manpower even in boom times. Furthermore, automation meant no job was secure. Modern manufacturing broke down production into a series of steps or tasks, and if a machine performed a task better or more cheaply, the worker lost his job. Automation and competition for work caused the unemployment rate in the second half of the nineteenth century often to run as high as thirty per cent. During the 1870s and 1880s strikes erupted, not orchestrated by labor unions but spontaneously born of frustration from dangerous—often appalling—working conditions, low wages, and ubiquitous layoffs (Trachtenberg 1982, 88).

From the end of the Civil War, American industry grew at such breakneck speed that by 1900 US industry outstripped England and Germany (Trachtenberg 1982, 52-3). In less than a century, about the length of a human lifetime, the United States had transformed itself from a rural democracy into the world's leading exponent of rambunctious capitalism. Not surprisingly, the rapid expansion of American manufacturing "came in hasty disorder" (Wiebe 1967, 20). The burgeoning economy was fraught with risks not only for the common laborer but also for the men who started up new enterprises. Every dip in the financial cycle brought a slew of business failures. The Stanley brothers were among the survivors.

The Stanleys shared with other entrepreneurial types of the late nineteenth century a near religious faith in science and a passion for the fruits of technology. They too used automation, but the Stanleys never valued machines over human intelligence. Wherever possible the brothers applied technology according to earlier standards of unit and small batch production. In contrast to the practice of most contemporary manufacturers who gave increasing authority to the plant manager for the sake of efficiency, the Stanleys stuck to the idea of maintaining quality with a hands-on job foreman. Their insistence on the highest standards of excellence explains in large part why they refused to take up the mania for mass advertising. While their business peers worked on public relations, the Stanleys relied on personal demonstration. The brothers never abandoned the belief that the customer needed to know what product he was buying, not who else was buying it (Trachtenberg 1982, 137). This adherence to old-fashioned Yankee craftsmanship would set them apart from the mainstream of the Gilded Age.

The Stanleys were by no means alone in their stubborn adherence to standards of an earlier day. Many ordinary Americans, while eagerly partaking of the new prosperity, felt bewildered by change and tried to carry old values into new circumstances. But the upheavals caused by industrialization were so great, the societal changes so profound that most efforts to apply the familiar to the strange simply failed. As historian Robert Wiebe wrote in *The Search for Order*:

> American institutions were still oriented toward a community life where family and church, education and press, professions and government, all largely found their meaning by the way they fit one another inside a town or a detached portion of a city (12).

In rapid succession the old neighborhoods and patterns of living that traditionally gave context to peoples' lives were being transformed by the growth of a national economy.

Nowhere did institutions seem more outmoded, and nowhere did people feel a stronger sense of dislocation than in the burgeoning cities of this period. Overcrowded, unsanitary, dangerous, and polyglot, they presented a set of problems Americans once believed were relegated to the Old World. Big city governments were so swamped by the speed and the extent of growth that for a time they could not deliver even the most rudimentary services to their inhabitants. City dwellers, most of whom were fearful strangers to one another, could not deal with garbage, sewage, fire and crime through familiar methods of community cooperation (Wiebe 1967, 14). In the new city, differences between rich and poor seemed greater and, for the first time in American experience, unbridgeable. The urban poor, many of whom were immigrants, turned for relief to the political ward boss who, in payment for loyal party support, supplied everything from turkeys for their Thanksgiving tables to public utilities for their streets.

The new industrial middle class responded to urban chaos largely by escaping it. Legions of managers, engineers, bankers, lawyers, doctors and businessmen, many of whom counted on the city for their livelihood, preferred to live in the exclusive and homogeneous safety of the suburbs (Schlereth 1990, 18). In 1890, the Stanley twins, flushed with the success of their first business venture, left Maine behind and settled in one such exemplary "garden city"—Newton, Massachusetts.

Like most of the major changes in nineteenth century American life, the shift of population to the borderlands began before the Civil War and gained momentum before the close of the century. As early as the 1830s a pattern of wealthy urbanites retreating to the country for the summer was well-established (Stilgoe 1988, 24-25). On the green lands beyond the city limits, gentlemen farmers grew fruit trees and bred livestock while their wives gardened with scientific fervor. For many this lifestyle became a permanent arrangement with the extension and improvement of rail service from big cities to the surrounding countryside. Trains made it possible for the rich to live on their borderland estates while commuting to the city. The relatively high cost of transit ensured that the urban poor could not follow.

In the second half of the nineteenth century, as the number of well-off Americans increased and the desirability of life in the city decreased, more of the new middle class chose not to live where they worked but instead joined the gentlemen farmers in suburbia. Despite their exodus, a number of these urban refugees dedicated themselves to salvaging the nation's cities. From the safety of their green redoubts, middle class reformers tried hard to reconfigure the new city into a place that was both congenial and aesthetically pleasing. Leadership for reform came primarily from the literary elements in American society; but for many members of the technocracy, infusing the new industrial order with a degree of social responsibility and "respect for cultural standards" was as important as making money. In Boston, perhaps more than in other American cities, the intelligentsia overlapped peacefully with successful capitalists to form a single reforming class (Howe 1976, 12).

The guiding rationale of urban reform was the belief that the influence of taste, culture, and natural beauty upon the masses would restore order to the nation's cities. Landscape architect, Frederick Law Olmsted, who was a prominent exponent of the city beautiful movement, maintained that a park would bring even the meanest individual to a sense of "courtesy, self-control, and temperance." Along with urban park systems such as Boston's vaunted "Emerald Necklace," came reconfigured downtown areas and grand new edifices. To the end of restoring peace and spreading enlightenment, an unprecedented number of monumental public buildings went up in American cities during the latter decades of the nineteenth and the early years of the twentieth centuries. Classical concert halls, civic opera houses, municipal libraries, and museums rose like fortresses. Among the most ardent proponents of this community uplift were suburban matrons recently freed from the drudgery of keeping house and eager to find new outlets for their energy. Both Stanley wives, Augusta and Flora, were enthusiastic patrons of the Boston Symphony, frequenters of art galleries, and diligent members of numerous clubs dedicated to redressing the ills of modern society.

Until the day, however, when culture again made cities habitable, the middle class technocracy lived contentedly in their suburbs. In these smaller, orderly communities it was possible to resolve many of the problems that made urban life so intolerable. Although working class folk lived in the suburbs too, these cities and towns became most notable for the comforting homogeneity they offered to the new middle class.

For people who had experienced breathtaking changes in their world, the opportunity to associate with others of similar means, to live among those in the same or like professions, offered affirmation of who they now were. Having grown up in a society in which a man lived where he worked and a woman worked all day just to keep the household going, the newly affluent had to adjust to a way of life in which disposable income and leisure time were the norm.

Amid the peaceful settings of suburbia, prospering technocrats and their wives evolved a social order that was at once self-conscious and public-spirited. They followed elaborate codes of behavior and exhibited great care in how they chose their friends and activities. Although status seeking informed many of those choices, it must also be noted that being seen in the right places required participation in the many civic ventures of the day. Indisputably, the upshot of all that volunteering was betterment in the quality of community life. The new middle class also displayed tireless dedication to intellectual self-improvement. While they often greeted the work of the hack and the master with equal fervor, it is hard to deny that those early suburbanites attained a level of conversance with literature, science and the arts largely unmatched by their middle-class counterparts today.

It was not sufficient for a successful man of the day merely to ply his chosen trade. He needed to prove himself by his leisure pursuits as well. The Stanley twins offer positive cases in point. In addition to their mechanical enterprises, each of them made violins—creditable instruments at that—in his spare time. They also proved to be skilled essayists ranging over topics as various as aeronautical engineering, the nature of work, socialism, the history of the violin, and the need for free world trade. Francis Stanley never attended university but proudly cited "constant home study" as the source of his education. Given the extent of his inventive prowess and the erudition of his leisure activities, who would disparage his claim?

For all the ways in which the Stanleys embodied the ethos of the times, they differed markedly from the richest and most industrially successful men of the Gilded Age. In the first place, the twins lived and worked in metropolitan Boston, the "Hub" of the early republic but by the late nineteenth century a city falling behind New York. By virtue of its better natural harbor and ready access to the interior of the nation through the Erie Canal, New York City had taken over the commercial and financial lead in America, eclipsing the Brahmins in all arenas save the literary. The stunning wealth and ostentation characteristic of the Gilded Age appeared most notably in New York City and in the summer playgrounds of its wealthiest citizens. The same period unfolded in Boston on a smaller, more decorous scale.

More importantly, the Stanleys themselves pursued neither success nor wealth with the single-mindedness of inventor-entrepreneurs like George Eastman and Henry Ford, whose genius consisted of anticipating and shaping consumer trends. The brothers competed in the marketplace with both Eastman and Ford, and by any financial yardstick fell considerably short of the achievements of these industrial titans. Viewed strictly in business terms, there scarcely seems reason to regard these twin inventors. But when one takes the full measure of their lives, the Stanleys emerge as worthy subjects. They entered the industrial maelstrom of the late nineteenth century on their own terms consciously tempering modern enterprise with the values of rural New England and the early Republic.

Chansonetta Stanley Emmons' circa 1898 photograph of the Stanley homestead in Kingfield suggests the cozy charm that captivated Mrs. F.E. Stanley. Photograph by Chansonetta Stanley Emmons.

Stanley Museum Archives.

CHAPTER 1

No Help from the Grown-Ups

WHEN THE STANLEY TWINS moved their business from Maine to Massachusetts in 1890, the Gilded Age was in full swing. The brothers themselves were just beginning to reach the top of their corporate form. The Stanley Dry Plate Company, the first of their joint ventures, was grossing $20,000 per month. In a decade, income would equal five times that amount. The twins were still seven years away from building the steam car that would make them famous. Even so, by the time the Stanleys opened for business in Watertown, Massachusetts, they had already traveled a great distance from home.

Freelan Oscar and Francis Edgar Stanley were born June 1, 1849, in the small western Maine town of Kingfield on the banks of the Carrabassett River. The town was so full of their relations that the saying went you couldn't toss a stone without hitting a Stanley. This was not surprising considering that the twins' great uncle Solomon Stanley was one of the founding settlers. Long after the twins' enterprises had taken them from their rural beginnings, a trip back to Kingfield had restorative power. As a middle-aged woman, Francis Stanley's wife Augusta liked to anticipate that familiar view as she motored into the town. She recorded one such homecoming in her diary on August 8, 1911: "...(T)hen up over the hill once more where we could see the old house. It is just as beautiful scenery as ever..."(Stanley Museum Archives).

The first English Stanleys appeared in New England around 1645. Great Grandfather Stanley moved his brood from Attleborough, Massachusetts, to Winthrop, Maine, during the Revolutionary War. Considering that he named his oldest son "Liberty," he must have been a patriot, not a Tory seeking safer ground. One of his other sons, Esquire

A panoramic view of Kingfield, population about 600, taken circa 1898 by Chansonetta Stanley Emmons. Mt. Abram (on the right) rises 4,000 feet above the town nestled in the Carrabassett River Valley. Photograph by Chansonetta Stanley Emmons.

Stanley Museum Archives.

Solomon Stanley (1780-1875), explored the territory of western Maine with William King, the man who became the first governor of the state. It was Solomon Stanley who chose the site and built the home that Augusta Stanley found so charming.

Solomon settled in Kingfield in 1808 and eventually acquired a 150-acre tract of land on both sides of the Carrabassett River (C. S. Emmons [1916] 1992, 6). He built a saw and grist mill and dye house in the town. Financial reverses forced him to give up his shares in the mills, but Stanley remained a leading member of the community.

Solomon's money woes seemed inconsequential alongside the difficulties incurred by his older brother Liberty (1775-1863). An inventor and violin maker, Liberty failed utterly at business. He could neither run the fulling mill his father had put him in charge of nor protect his inventions, which included a cloth-shearing device. When his young wife died in childbed, the blow so overwhelmed Liberty that he parceled out his seven children to relatives and neighbors. The youngest of the orphans was taken in by his uncle and namesake in Kingfield, Solomon Stanley. Childless himself, the elder Solomon raised Liberty's boy as his own and imparted to him the habit of civic responsibility.

Young Solomon (1813-1889) grew up to be a man who revered learning. He both taught school and farmed the family homestead. Demonstrating the Stanley flare for stringed instruments, he made violins and played the bass viol, even in church where the practice was frowned on among Congregationalists. In 1830 when one of the many waves of temperance reform swept the country, he took a teetotaler's vow, a pledge that strongly influenced his twin sons. Solomon Stanley served frequently as town selectman and was even elected to the state legislature although he was a Republican in a Democratic town.

As a Kingfield selectman, Solomon II revealed a talent for behind-the-scenes maneuvering. For example, he supported the idea of building a graded schoolhouse at Kingfield but saw the plan voted down repeatedly in town meeting. When the issue came up for reconsideration in 1872, the town meeting adjourned before a vote could be taken. The following spring, Stanley remembered that adjournment had left the schoolhouse issue tabled but alive. For the upcoming meeting, he quietly assembled a majority of citizens in favor of building the new school, and when the matter came to a vote, Stanley had the forces to defeat the usual gaggle of resisters (C. S. Emmons [1916] 1992, 12). Solomon Stanley's attention to detail and cracker barrel cunning apparently rubbed off on his famous twin sons, who displayed these traits as the signature style of their business dealings—though not always to the same good effect.

Even Solomon's shrewdness, however, did not extend to financial matters. An ill-fated partnership in a general store left him burdened with debt throughout much of his life. Despite his lack of worldly success, Solomon Stanley remained an important man in his little community. Testament to his standing in town affairs was the fact that the Stanley homestead became a prominent landmark in Kingfield. The family pasture land was known as Stanley Hill and the farm stream, dotted with waterwheels and other contrivances rigged by Solomon's twin sons, was called Stanley Brook.

Isaac Newton Stanley, born in 1841, was the oldest child of Solomon II and his wife, Apphia Kezar French (1819-1874). Miss French was herself the daughter of a founding Kingfield family and may have been the source of the theatrical streak that surfaced intermittently in future Stanley generations. If not the originator of the family taste for dramatic appellation, she clearly threw herself into the spirit, mining not only family history but science, literature, and religion to find appropriate names for her seven children. Isaac Newton was not the first Stanley to be named for the great scientist. Liberty had chosen that name for his oldest son. Isaac II distinguished himself as the only offspring of Solomon and Apphia to fight in the Civil War. Apparently his stint in the Army of the Republic was enough drama for one lifetime. Isaac came home to live out his days on the Kingfield homestead.

The twins, Freelan Oscar (Freel) and Francis Edgar (Frank) were born eight years after Isaac. Two boys followed who died in young manhood of tuberculosis. John Calvin (1852-1883) was named for his

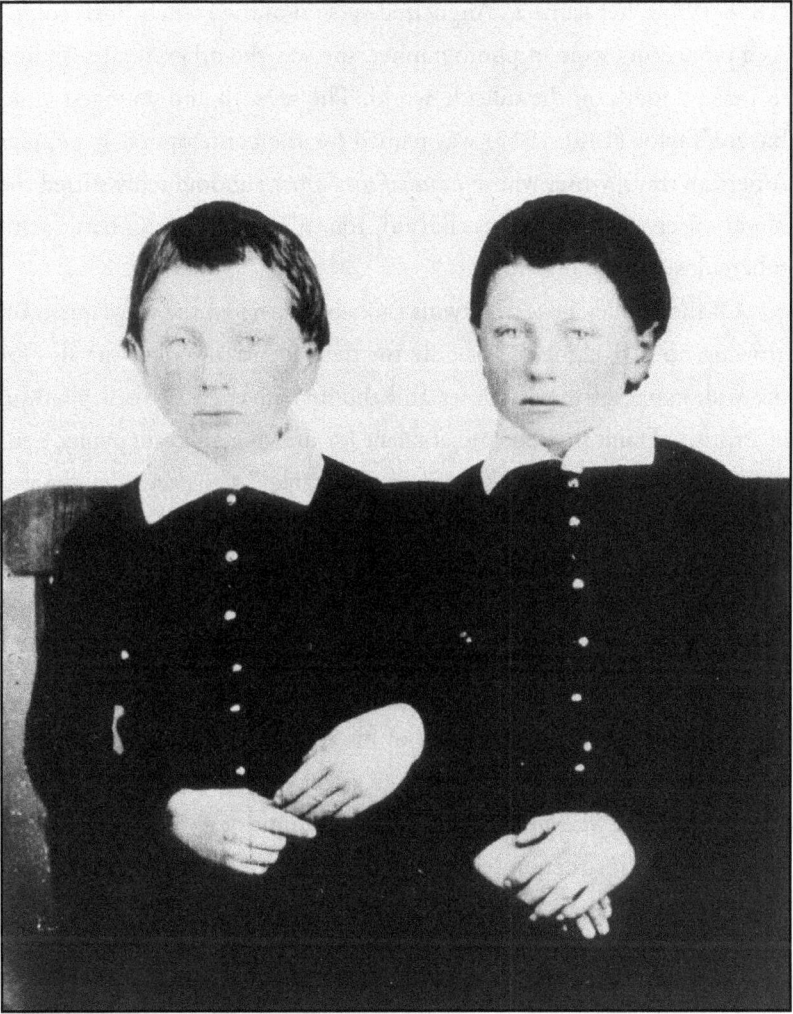

The Stanley Twins, Inc., around nine years old – the age when they launched their first mechanical enterprises.

Stanley Museum Archives.

maternal uncle, John Calvin French, who in turned had been named in honor of the great Protestant divine. Solomon Liberty (1854-1881) was the namesake of both his adopted and biological grandfathers, the latter of whom was noteworthy at least as the source of the family tinkering gene. The only girl in the family was called Chansonetta

(1858-1937), her name an Anglicized version of the French "little song." As a pioneering woman photographer, she was the other Stanley sibling to make a mark on the outside world. The seventh and youngest child Bayard Taylor (1861-1915) was named for the contemporarily popular American travel writer whose *Land of the Saracen* undoubtedly stirred the blood on cold Maine nights. Bayard, too, died after a long battle with tuberculosis at age fifty-four.

Of the Stanley boys, the twins probably garnered the most attention growing up in Kingfield, not only for their identical looks, but also for the wide range of their abilities and the inventiveness of their youthful enterprises. Frank showed special talent for drawing and calligraphy, penning what would become the trademark "Stanley" signature when he was just fifteen. Freel carried on the family love of stringed instruments by making a violin for himself when he was eleven. At sixteen, he fashioned an instrument—still extant and on display in the Stanley Museum—that Maine violin maker Jonathan Cooper describes as "playable" and "showing talent" (Cooper 1986, 6). To have created such an instrument is no mean accomplishment, both on account of his youth and because Freel had so little to go on. There were no master violin craftsmen to learn from in America as there were in Europe, only the example of factory-made instruments and the instruction of his father, who had learned from his father.

The twins began their mechanical projects at an even earlier date (F.O. Stanley [1930] 1991, 14). By age nine they had rigged a little mill at the site of a steep drop-off between two ponds. As Freel would later

The Stanley twins' signatures as fixed on a 1901 joint patent application.
U.S. Patent and Trademark Office.

Apphia Kezar French. Daughter of Kingfield pioneers and mother of the Stanley twins.

Stanley Museum Archives.

relate, their plant, constructed "without any help from the grown-ups," consisted of both a turning lathe and a waterwheel to power it. "The waterwheel," Stanley wrote, "had a vertical shaft, but a turning lathe has to have a horizontal shaft. This necessitated beveled gears which we also made." At first the brothers manufactured tops for a penny apiece but soon exhausted the market. They had considerably more success with weaving spools. In 1859 Kingfield housewives still produced most of their own woolen cloth with a spinning wheel and handloom. The seven-inch spools the twins manufactured and sold for two cents each were part of the warping bar on the loom. Since there were 150 spools on each bar, the young Stanleys had hit upon a product every housewife needed in large quantity.

With the profits from another of their youthful enterprises, the Stanleys bought themselves a copy of Greenleaf's *Introduction to the National Arithmetic*—a standard text of its day—and gained a measure of regional fame as the boys who had ciphered all the way through Greenleaf's. As they grew older, their obvious intelligence and family influence inclined the Stanleys towards a teaching career. By the time they came of age, the Civil War was over and the country was embarking energetically on an era of professionalism. Even in a rural community, it was often no longer enough for the best pupil in class to take over from the local school master; the idea was gaining ground that teachers train for their work at a professional school. Thus in 1869 both brothers set out to study at the newly opened Western State Normal School at Farmington, some twenty miles south of Kingfield.

Frank's student days at Farmington were notable only for their brevity. It is not clear how long he remained, but he left school without graduating, incensed because an instructor doubted his word that he had drawn a map freehand (Davis 1997, 14-15). That incident apparently soured his enthusiasm for formal education altogether. As an older man, asked to provide biographical information for a book on the leading citizens of Newton, Massachusetts, Stanley described himself as self-taught (Brimblecom ed 1914, 215).

Frank Stanley and his "Gustie" at the time of their marriage in 1870.
Stanley Museum Archives.

Freel followed a more conventional track, graduating from Farmington in the Class of 1871. He then studied at Hebron Academy near Auburn, Maine, before enrolling at Bowdoin College. At the time he began his studies at Bowdoin in 1873, the college required under-graduates to complete a three-year program in military tactics along with the regular curriculum. Memory of the Civil War remained fresh, and Bowdoin's president had been one of Maine's most illustrious soldiers, having fought in twenty-four battles and sustained six wounds. Before the war, Joshua Chamberlain taught rhetoric at Bowdoin and returned afterwards in triumph as its president. His introduction into the curriculum of courses in modern science and engineering no doubt caught Freel's interest, but Chamberlain's insistence on military training grated on the independent-minded Stanley.

During Freel's first term at Bowdoin, President Chamberlain additionally required each man to purchase a three-dollar uniform. Student grumbling over the college's military requirements gathered into campus revolt. Chamberlain brought the matter under control by threatening to expel any student who did not buy a uniform and sign a loyalty pledge at the start of the new school year. Freelan Stanley was one of just three undergraduates who chose not to return in the fall (Pickering 2000, 13-15). Many years later when asked by his nephew Raymond why he had not finished college, Freel dryly noted that he couldn't afford the cost of a uniform.

Interestingly, both Stanleys relinquished the chance for college degrees just when professional training was beginning to be prized. One can attribute their actions to the rashness of youth, but it is possible the brothers left school because they were already confident that they could learn on their own. What looked like adolescent impetuosity may have harked back to the determination of a pair of nine-year-olds to construct a turning lathe "without any help from the grownups." It clearly presaged a lifetime of intellectual self-sufficiency. In all their adult endeavors, the Stanleys proceeded with the conviction that they could master what they needed to know and solve any problem by their own industry.

Of the two, it is probably fair to say that Frank, who quit school for good at the first challenge to his integrity, had the harder head and the more independent spirit. Perhaps Freel's weaker constitution (he endured frequent bouts of tuberculosis) played a part in his personality development, but throughout their lives Frank took the bolder steps and, until the twins formed a business partnership, enjoyed the greater success. Late in his own life, Frank's son Raymond recalled his father as the more energetic one of the pair: "All his life my father was very, very active. He was very different from his brother who was very quiet. When you talked to Freelan, he'd close his eyes on you" (R. Stanley 1983, 4).

Despite quitting his professional training before it began, Frank had no trouble getting hired as a teacher. The town of North New Portland, a few miles southeast of Kingfield, took him on as both principal and teacher. Stanley had apparently taught in this town briefly before attending Farmington. In 1869, he returned to take up his dual posts and was later joined by his fiancé, Augusta May Walker. Augusta was a young woman who had acceded to the teaching profession the old-fashioned way—at thirteen she knew more than her teacher. Frank wed his "Gustie" on January 1, 1870, and by the following December 14, the couple were parents of their first child, Blanche May. Despite the responsibilities of marriage and fatherhood, Frank did not settle into a rut. After several more years and several different teaching posts—including a stint as assistant superintendent at the Maine State Reformatory in South Portland—he was restlessly at work on a new career. He thought he might read the law, and, in order to finance his legal studies, Frank took up a boyhood hobby in earnest.

While at a position in Strong, Maine, another little community to the south and west of Kingfield, Frank renewed his passion for drawing and sketching with a series of self-portraits, none of which is extant. The medium he chose to work in was liquid black crayon. His daughter Blanche recalled her father teaching himself the technique: "His first

In this crayon portrait of his four-year-old daughter Blanche, Frank Stanley used his atomizer and a piece of lace as a stencil to spray on the collar of her dress. Portrait by F.E. Stanley.
Augusta Tapley Collection, Stanley Museum Archives.

Photographic portrait of the artist in his Lisbon Street studio in Lewiston.
Photograph by F.E. Stanley.

Stanley Museum Archives.

attempts were very black. As soon as he finished one he would tear it up, put it in the stove, burn it, and start all over again" (B. Hallett 1954). As he began to master the medium, he sought subjects other than himself. He rendered his work in a realistic style, uncompromising in detail. Stanley's portraits, like the work of the superb American artist Thomas Eakins (1844-1916), bore the impress of the new and increasingly popular art of photography.

By 1874, Stanley had begun to travel some distance to peddle his portraiture in the Lewiston-Auburn area northwest of Portland. When he knocked on the door of the editor of the *Lewiston Journal,* Frank found enthusiastic patrons and lifelong friends in Nelson Dingley and his younger brother Frank, who owned and managed the paper together. Nelson Dingley had an influential editorial voice amplified by long service in the Maine State Legislature and one term in the governor's office. He and his brother remained close to Stanley even after the latter moved his business to Massachusetts, and Nelson went on to serve many years in the US Congress (David C. Young and the Androscoggin Historical Society).

As Frank got more involved with crayon portraiture, he grew impatient with the slowness of the process of application. Soon he was experimenting with the use of an atomizer to speed his strokes. From this simple step he got the inspiration for his first invention. On September 19, 1876, Stanley was issued U.S. patent number 182,389 for the substantial improvements he had made on the basic atomizer. He called his invention the airbrush—the same tool used today primarily for photographic touch-ups. Stanley modified the atomizer so that he could produce a fine line or a broad brush stroke by changing nozzles. The refinement of control Stanley built into his airbrush let him achieve a variety of portrait effects from shading the brow line to drawing single strands of hair. He found the airbrush extremely useful in carrying him over what he considered the tedious parts of painting. To render the lace collar on the dress of his five-year-old daughter, for example, he used an actual piece of the lace as a stencil and sprayed the pattern onto the canvas (Davis 1992, 13). Both Stanley brothers later found the airbrush a useful tool for varnishing violins.

As he had hoped, the airbrush significantly speeded up the process of creation and Stanley began turning out crayon portraits of prominent Maine figures like Nelson Dingley. Modeling at times from photographs, Frank produced portraits of such venerable Americans as Henry Wadsworth Longfellow and William Cullen Bryant. According to Stanley family lore, F.E.'s portrait of James G. Blaine, the powerful Republican Senator from Maine, took first prize at an exhibit in Portland and may have been part of the Maine State exhibit at the nation's 1876 Centennial Exposition in Philadelphia.

Along with artistic recognition, monetary prospects were picking up for Stanley as well. At the urging of the Dingleys, he put up a sign in a local shopkeeper's window advertising his work. Soon Stanley had so many orders to fill that he was able to open his own studio in Auburn. Fortunately, Stanley's growing business had kept pace with his family; a second daughter, Emma Frances, was born August 15, 1876. Despite having two little girls to tend, Augusta got actively involved in the financial end of the business when she discovered that Frank was a bit dilatory about billing. In order to work in the office and mind her children, Gustie created a day care room right at the studio (B. Hallett 1954).

By 1878, Frank had moved his business to nearby Lewiston. Stanley's Lisbon Street studio, large and elegantly appointed with his work on prominent display, readily attracted the carriage trade. It became particularly fashionable to have a Stanley portrait of one's children. A well-known story about Frank's ability to capture a likeness involves a portrait he painted for the grieving parents of a dead child. With only their description to go on and the surviving children as models, Stanley is said to have produced a startlingly accurate portrait.

The growing volume of his business had Stanley again thinking up ways to save time. The idea came to him, perhaps because he had already worked from photographs, to hire an assistant to take a picture of each client. From the photographic image, Stanley then painted the finished crayon portrait. This procedure cut down significantly on the length of sitting (or squirming) time for the client. But Stanley was not fully satisfied with the results because the quality of his work was now contingent on another man's skill. Dissatisfaction led to the purchase of Stanley's first camera.

Six years after his brother married, Freelan wed Flora Jane Record Tileston.
Independent-minded Flora often ran the Cambridge shop while Freel struggled to rebuild their factory in Maine. Flora's photo by F.E. Stanley; Freelan's photo by Curtis.
Stanley Museum Archives.

CHAPTER 2

Stanley Brothers Incorporated

PHOTOGRAPHY WAS IN ITS INFANT STAGES at this point, having first been introduced to the world in 1839 by Louis Daguerre who discovered that silvered copper sheets had a photosensitivity that made it possible to imprint pictures of objects. The popularity of the daguerreotype in the United States was overwhelming. Something in the matter-of-fact nature of the medium seemed to touch a chord with the practical American soul. Ordinary folk, who might never have a chance or even the inclination to sit for a painted portrait, were willing to oblige the camera a moment for posterity.

In the early days of photography, only professionals and a handful of amateurs took pictures. Since there was no mass market for cameras or photographic accessories, improvements in the science of picture taking occurred not from the press of economics but the spirit of inquiry. European scientists in the fields of chemistry and optics produced the greatest advances in photosensitive material. Experiments with "gun cotton," raw cotton treated with nitric and sulfuric acids, revealed that it was soluble in sulfuric ether. The resultant sticky, transparent substance was called collodion. When applied wet onto glass, it had greater light sensitivity than daguerreotype, was cheaper and could make multiple prints. In short order, collodion replaced daguerreotype as the photosensitive material of choice. An even greater advance occurred during the 1870s when Richard Maddox, an English doctor, developed the first dry collodion by passing silver-halogen salts through gelatin. Maddox's formula proved much more light sensitive, with considerably longer viability, than the wet collodion. Dry plate, as it came to be called, created a revolution in the photographic industry. For the first time, gelatin coated glass plates could be mass-produced (Jenkins 1975, 1, 5, 10). To photographers, dry plate meant quicker exposure times and precluded the need to carry portable darkrooms around with them.

Throughout the 1870s, European companies turned out the best quality dry plate, while in America business was confined to distribution. Photographic jobbers or wholesale merchants ordered dry plates from abroad, then sold the merchandise to photographers across the country. It was only a matter of time, however, before American manufacturers turned to dry plate production. The nation's deep labor pool and tariff protection policy for infant industries all but guaranteed success to the company that could put out a product comparable to or better than European dry plate. By the end of the decade, the three largest photographic jobbing houses were trying to do just that.

The first Americans to successfully offer native photographers a creditable alternative, however, were not in the distribution business but tended, like Gustav Cramer of St. Louis, to be men with chemistry backgrounds (Jenkins 1975, 12). Yet, the man who turned the American dry plate industry into a fast-paced, competitive market had no scientific credentials at all. Bank clerk George Eastman of Rochester, New York, decided it would not do just to produce factory dry plate; he sought to automate the process and in 1880 got a patent on a plate-coating machine. Eastman built his prototype in 1877, and by June of 1879 he was in London, then the center of photographic science, to procure a British patent and sell the rights. He returned to America before the patent came through but still managed to interest two English dry plate manufacturers in his machine (Brayer 1996, 31-32). The stir caused by Eastman's patent helped launch him into business. In the long run, however, that initial boost proved more valuable than the machine itself, which looked better on paper than it ever worked in the factory. The invention of a successful coating machine would not occur until six years later as the brainchild of a pair of Yankee entrepreneurs.

During the early 1880s, a prospering domestic dry plate industry spawned numerous start-up companies—most of which failed because they could not obtain the necessary scientific information, or produce on a scale sufficient to compete, or land a sole agency with a jobber (Jenkins 1975, 74). Defying these odds was a young Maine portrait artist recently launched into the picture-taking business and now dissatisfied with the quality of the dry plate he used.

Frank Stanley went about learning how to make dry collodion the way he went about everything—on his own. He read the scientific journals and began experimenting with a formula. By 1880, he was using plates of his own making and selling them locally to other photographers. His product came to the notice of C.H. Codman & Company, a Boston-based photographic stock house whose representative came to Lewiston to urge Stanley to manufacture his plates and to allow Codman to peddle them. With an initial investment of $500 Frank Stanley went into production (F.E. Stanley [1914] 1991, 12). In time his Lewiston portrait studio included a dry plate factory on its lower floors. But he did not get to that point of prosperity alone. Soon after going into dry plate manufacture, Frank Stanley took on a familiar business partner.

Brother Freel had left Bowdoin College to become a teacher. He taught in public schools in Maine and Pennsylvania before returning in 1880 to the Normal School at Farmington as an instructor. Failing health—his first brush with recurring tuberculosis—forced him to resign that post after just one year (Parrington 1889, 68). When he regained his health, he saw the need to "go into some less sedentary occupation." Thus Freel left teaching for manufacturing. His interest in practical education, however, remained keen. While at Farmington, he had written an essay urging the use of "apparatus" for a hands-on method of teaching geometry. He himself came up with the idea of manufacturing drafting tools for students.

Stanley packaged a metal compass, a paperboard ruler and protractor, a sheet metal triangle and right angle in one handy wooden case.

"Stanley's Practical Drawing Sets," as he called them, sold for $1.00 apiece. By the early 1880s, Freel had a factory set up and running in Mechanic Falls, Maine. But his enterprise did not thrive for long.

Fire put him out of business and into debt by 1882. With a wife to consider (he had married teacher Flora Jane Record Tileston on April 18, 1876) and the family history of business failure hanging over him, Freel moved quickly to save his venture. He took a temporary job as assistant cashier in his friend G.L. Damon's American Steam Safe Works in Cambridge, Massachusetts, and borrowed money from Frank to get back on his feet. Freel and Flora rented an apartment in nearby Somerville but optimistically kept their home in Mechanic Falls. They opened a temporary shop at 17 Main Street in the Steam Safe Works until their factory in Maine was ready for business. Flora herself worked in the

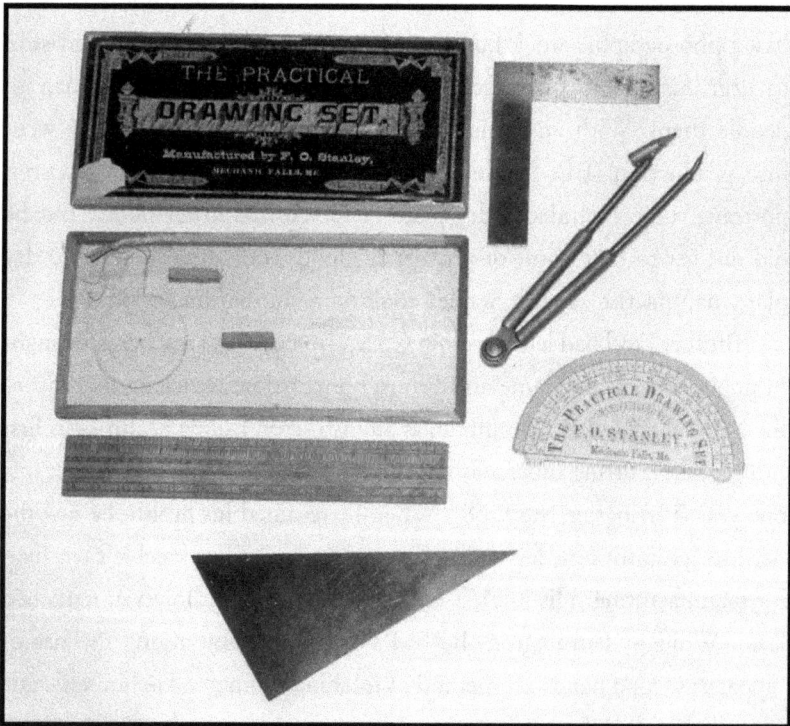

The Stanley Practical Drawing Set: a compass, ruler, protractor, right triangle and square, with carrying case. Photograph by John S. Dillon.
Courtesy John S. Dillon.

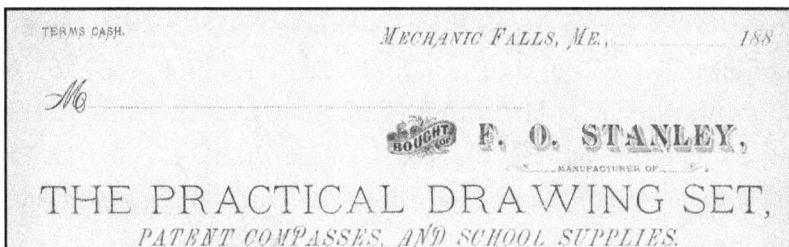

A business bill head from F.O. Stanley's Mechanic Falls shop.

Stanley Museum Archives.

Cambridge shop and ran it alone whenever Freel traveled to Mechanic Falls.

Even as Freel attempted to recover his business, Brother Frank had a proposition for him in Lewiston. Having had no luck with two plant managers in almost as many years, Frank hoped to convince his brother to join him in the dry plate business. But the Stanley sense of independence, coupled with the natural rivalry that must have existed between two such close and gifted individuals, probably kept Freel from jumping to accept Frank's offer. From May throughout the summer of 1884, Freel mulled over the prospect of partnering with his brother. In the end, Frank's invitation proved too tempting for Freel, whose business recovery seemed increasingly doubtful. According to Flora's diary, he had decided to accept the offer by the end of August. On the twenty-eighth of that month she noted that "F.O. went to the shop, packed drawing sets, and started for Lewiston."(Stanley Museum Archives). She herself followed a few days later, and the couple was soon looking for a place to live.

The official "partnership business commenced" on October 27, 1884, when the Stanley twins, who had last worked together in the warping spool business, once again joined forces. By early November, Flora and Freel had given up their home in Mechanic Falls and moved to 141 Pine Street in Lewiston. A month later, Flora was dispatched to Cambridge to oversee the dismantling of the drawing tool business.

How happy Flora was with Freel's decision to move to Lewiston is not known. It seems doubtful, however, that she looked forward to living in the same town with her sister-in-law. She and Augusta did not share the bond of their twin husbands. Indeed, for people drawn so closely together in life by circumstances, they seemed to avoid each other's company whenever they could. Gustie, the more prolific diarist, often wrote patronizingly of Flora with phrases such as "She looked good for her" and "Flora was even more reticent, if that is possible." For her part, Flora put the case directly. Shortly after settling in Lewiston, she wrote in her diary on November 29, 1884, that Frank's family had paid them a housewarming call: "Gustie," Flora noted, "was as amiable as a meat-ax" (Stanley Museum Archives).

At the age of eighty-six, Freel (known in adulthood as F.O.) wrote a history of this first grown-up business venture with his brother. "I doubt if an enterprise of any considerable size was ever started in a more primitive manner than we started the manufacture of photographic dry plate" (F.O. Stanley [1936] 1987, 14). Noting that they had "practically no capital, no factory, and no customers," he attributed the success of their business to making a product cheaper in price but equal in quality to their competitors' plates. Professional photographers of the day, who mostly did portrait work, wanted dry plate that did not require a long exposure time. According to F.O.'s account, within two months of their partnership, the brothers were turning out dry plate that was twice as fast acting as the best plates on the market.

F.O. attributed their running start-up in no shy measure to his own efforts. Granting that Brother Frank performed the necessary task of "quality control" in his photographic studio, F.O. made it clear who worked on the loading dock (F.O. Stanley [1936] 1987, 8):

> I was the general manager, treasurer, bookkeeper, and laborer. I never struck for increased wages or shorter hours, although I worked an average of more than 13 hours a day. I washed the glass, made the emulsion, coated the plates, and when dry, packed them into the paper boxes, and the boxes into [shipping] cases...

In the earliest days of the business, F.O. no doubt worked harder than his brother did because the former came to the partnership in debt.

Show them and they will buy. Frank Stanley vaults a puddle to demonstrate the speedy action of Stanley dry plates.

Stanley Museum Archives.

By the following spring, however, business was so good that Freel saw himself "on the road to wealth." In a letter dated May 7, 1885, he wrote to Flora, who had been visiting in Mechanic Falls, that he was going to have "all my debts paid the first of June." Just one week later, he sent her an enthusiastic update, "We have enough to pay our debts and leave us more than before the fire" (Stanley Museum Archives).

The early success of Stanley Dry Plate seems even more impressive considering that the brothers launched their enterprise during the recession of 1883-84 and outlasted a welter of competition. Many start-up dry plate companies foundered on their inability to secure an exclusive jobbing contract, but the Stanleys got along surprisingly well by selling directly to photographers. By F.O.'s account, some of the most successful practitioners in New England bought Stanley plates. This included the acclaimed William Notman, who had studios in New York, Boston, and Albany, in addition to his headquarters in Montreal. An advertisement for Notman's photography in 1889 listed Stanley plates for sale at his studios (Notman Photographic Archives).

The Stanleys never advertised in the trade papers—a fact that F.O. relished in his account—but preferred to convince potential dealers and customers by demonstrating the superiority of their product. F.O. claimed that one such personal test led to the brothers' garnering a sole agency with E. & H.T. Anthony, the largest photographic wholesaler in the country. By Freel's telling, Brother Frank had gone to Anthony headquarters to convince Colonel Wilcox, the company general manager, to carry Stanley plates. In the midst of his sales pitch, William Notman arrived fortuitously to see Wilcox. Apparently Notman launched into an unsolicited endorsement of Stanley plates that convinced Wilcox to carry the line (F.O. Stanley [1936] 1987, 16). Shortly, thereafter, Wilcox himself journeyed to Lewiston to offer the brothers sole representation. E. & H.T. Anthony indeed must have been impressed with the quality of Stanley dry plate. But the immediate motivation for the deal was probably loss of the clientship of George Eastman, who had broken his contract with Anthony in 1885 because he was already more interested in roll film (Jenkins 1975, 76).

The Stanleys themselves soon quit their deal with Anthony. F.O. explained that they withdrew because this exclusive arrangement antagonized other photographic dealers who then urged their clients to use competitors' plates. It seems just as likely that the Stanleys, having developed an in-house marketing department, simply outgrew the need for an exclusive jobber (Jenkins 1975, 76). Between 1887 and 1889, Freel was one of three agents who took to the road with the Stanley product. Sometimes traveling for two months at a time, he covered territory as far west as Colorado, as far south as Kentucky, and as far north as Minnesota. He wrote letters to Flora from every hotel stop: Buffalo, Cleveland, Lexington, St. Paul, Kansas City, Chicago, Cincinnati, Rochester, Albany, Philadelphia, St. Louis, Detroit, Omaha, Toledo, Indianapolis, Syracuse, Denver, and Milwaukee.

The Stanleys may also have gained confidence from a manufacturing innovation they came up with in 1886 that guaranteed their profits would continue to rise. George Eastman had been the first to patent an automatic plate-coating machine—some six years before the Stanleys—but the Stanleys were the first to invent one that worked. Eastman had the temerity and luck to convince investors of the viability of his invention before its effectiveness was proved. But the account of his first plant manager, Florence S. Glaser, reveals that Eastman's device was a dud never used in the actual production of Eastman dry plate (Brayer 1996, quoted p.34):

> ...(T)his coating machine in any of its various versions never did work right...it was hard to clean after a run, it was wasteful of the expensive emulsion, and it was always likely to cause bubbles, waves, and streaks on the plates—all this despite Eastman's claims to the contrary.

By contrast, when the Stanleys decided to build a coating machine it was to gain a competitive edge in an already crowded dry plate market, and they made sure their invention turned out a product superior to hand-coating. F.O. claims to have had the seminal insight that plates could be coated "the way paper is made," then to have set about building a machine to carry the plates and chill them. Frank (F.E.) designed the device for spreading the emulsion over the plates. By F.O.'s calculation, the brothers began automatically coating their plates within thirty days from drafting board to prototype. They were issued the first of their many joint patents on July 13, 1886.

Normally it took two workers—the Stanleys employed young women for this task as well—one minute to hand coat a single 8x10" glass plate. The Stanley machine increased productivity from sixty plates per hour to sixty plates per minute with just one man working. Not content with merely speeding up the process of application, they invented a washing machine to prepare the glass plates for coating. Their contraption soaked the glass in dilute sulfuric acid, scrubbed both sides of the plate with brushes, rinsed both sides and applied substratum to the side that received the emulsion. Working by hand, a washer and his helper could ready sixty such plates for coating in one hour. The Stanley washer increased output to sixty per minute and eliminated one worker from the process. Although the brothers included the glass washing apparatus as part of their patent, this portion of their petition was denied on the grounds that it too closely resembled a recently patented dish washing machine.

The Stanleys' automation, which speeded up and streamlined the most labor-intensive parts of dry plate manufacture, cut the cost of production to one-sixteenth the cost incurred by their competitors (F.O. Stanley [1936] 1987, 9). This enabled the brothers to halve their retail price from the standard $4.80 per dozen to $2.40. According to F.O., "it is difficult to describe the storm that this created." He recounts in his folksy style how Gustav Cramer of St. Louis arrived at the Lewiston factory with a copy of the Stanley patent in his pocket. Convinced by a side by side comparison of his plates with the Stanleys' machine-coated ones, Cramer offered the brothrs $2500 on the barrel head to install their machinery in his factory. F.O. notes with amusement that it would have cost them "no more than $250" to ship their machinery to St. Louis, but the brothers rejected the offer because they wanted to retain their monopoly on dry plate coating (F.O. Stanley [1936] 1987, 9).

Perhaps they should have taken Cramer's money. Whatever market edge they gained lasted a brief three months. As F.O. explained, their patent claims proved too narrow to prevent competitors from easily building "machines that did not infringe on the (Stanley) patent." True to Stanley form, the twins had drafted this first joint patent application without any help from the lawyers. Although F.O. asserted that an attorney reviewed their efforts and found no fault with their work, thereafter the brothers let experienced patent lawyers draw up their petitions.

While the Stanleys' invention did not corner the market in automated production, their innovative burst did ensure the survival of their company and establish their reputation. On December 26, 1886, the Stanley Dry Plate Company was incorporated under the laws of Maine.

Two years later in February of 1888, the brothers expanded their operation into Canada (Notman Photographic Archives). They opened an office and factory—eventually run by the Stanleys' nephew Carleton—at 613 Lagauchetiere Street in Montreal. The decision to move north seems to have been influenced by William Notman himself, who had grown tired of importing dry plates to his Montreal headquarters. According to the records of the Notman Photographic Archives, he first attempted to convince a British dry plate company to set up a Canadian subsidiary. When this effort failed, Notman turned to the Stanleys. The timing was perfect; the Stanleys' business was thriving and their confidence high.

Within four years of incorporation and two years after Canadian expansion, the brothers were grossing five times more income than they had made before automation. Now they planned to increase profits even more by relocating their US factory to metropolitan Boston. The move south would eliminate a costly, inefficient arrangement whereby glass ordered from abroad arrived at Boston, was shipped to Lewiston, and sent back to Boston for distribution as finished dry plates.

The Stanleys were in their forty-first year when they began to net the kind of profits that would make them wealthy men. They already led lives dramatically different from those of their parents and male siblings. F. E. had earned recognition as one of the leading businessmen of Lewiston, and both brothers had taken up the leisure hobby of owning and racing trotting horses. Though not yet on a par with the Frank Dingleys, whose red brick mansion made them, in young Blanche Stanley's estimation, the "royalty" in town, the twins would soon outstrip their old friends on an even larger stage.

On June 4, 1890, eighteen tons of dry plate machinery shipped south from Lewiston to Boston. The brothers themselves soon followed. Still boyishly lean at mid-life and sporting identically trimmed beards, the twins assumed the sartorial style that became their signature. For the first time in their adult lives they bought matching suits of clothing (F. Stanley Diary, Nov. 8, 1890). With that dash of showmanship, the Stanleys set out to prove that country tinkerers could move comfortably into the ranks of Boston's elite technocracy.

CHAPTER 3

Happy is the Man

LIKE SO MANY OF THEIR PEERS, the Stanleys did not settle in the big city. They preferred to set up both home and shop in the suburbs of Boston, building their houses in Newton and their dry plate factory in neighboring Watertown. They may have chosen to live in Newton—a smaller city some ten miles west of Boston on the Charles River—because it was widely reputed for its beauty and civic amenities. Indeed, in the years since the Boston and Worcester Railroad came through its precincts, Newton had become a paradigm of the leafy suburbs that were increasingly home to America's new middle class.

Frank, Augusta, and thirteen-year-old-Emma moved to Newton on March 20, 1890, and every year thereafter Gustie noted the anniversary in her diary. Blanche Stanley, now nineteen, had preceded her parents to metropolitan Boston to study at the Cowles Art School. She had been living with relatives in the area but rejoined her family in their first home at 31 Jefferson Street. The house stood just inside the Newton line within easy walking distance of the dry plate factory, but as Blanche noted later, Jefferson Street while "eminently respectable" was "not really fashionable" (B. Hallett 1954). Two years later Frank built a home at a more suitable address, number 61 Franklin Street in Newton Corner (today number 261). Here the Stanleys' third child, and only son, Raymond was born on April 1, 1894.

For Flora and Freel, always more tentative than the F.E. Stanleys, the move to Massachusetts proceeded in stages. Flora was still living at 44 Pierce Street in Lewiston when Freel wrote to her on June 13, 1890: "The house we are to move into has been occupied for eight years by Mr. Elliott the Bicycle Man—and I think they (sic) must be the extreme of neatness" (Stanley Museum Archives). The Elliott house stood at 48 Maple Street in Watertown, with the Hickory Bicycle factory and the new Stanley Dry Plate headquarters situated behind it. Sterling Elliott, an avid and successful inventor, soon became friend, as well as landlord, to F.O. Their mutual love of mechanics and trotting horses sealed a lifelong friendship. Childless himself, F.O. named one of his prize trotters after Elliott's son Harmon (Elliott 1945, 25). Even after the F.O. Stanleys moved in 1894 to their own handsome residence on Newton's Hunnewell Avenue, they remained close to the Elliotts.

The city the brothers chose for their new home was a world apart from little Kingfield and incomparably more fashionable than Lewiston. But it had not always been so. Long before the railroad came through, Newton was a mixed farming and manufacturing community. Bay colonists first explored this inland area in 1631 and came to stay in 1639. Briefly part of Watertown, Newton was for most of its first forty-nine years annexed to Cambridge. One of the few settlements in the Massachusetts Bay Colony to be spared the bloody conflicts of King

Sterling Elliott, inventor and mechanical master, was F.O. Stanley's lifelong friend. Elliott manufactured bicycles in Watertown and later became a leading force in the "Good Roads Movement."

Stanley Museum Archives.

When they first moved to Massachusetts Freel and Flora rented this house, owned and built by Sterling Elliott, at 24 Maple Street in Watertown.
Stanley Museum Archives.

Philip's War (1674-75), Newton remained undisturbed—except for revolution against the King of England—for nearly two hundred years.

The first English settler was John Jackson whose farm bordered the Charles River near what is now the Brighton section of Boston. Jackson's homestead and a small clutch of farms south of the Charles River became known as Cambridge Village. Officially part of the town of Cambridge, these families had to cross the Charles in all weather and trek to the meetinghouse near Cambridge Common for Sunday services. Their vociferous complaints finally led to permission from the General Court of Massachusetts Bay to build a meetinghouse and lay out a burial ground on the south side of the river. Complete independence from Cambridge came soon after, and by 1688, Newtown (the "w" later dropped from use) was incorporated with a population of sixty-five freemen.

Since church attendance was mandatory in Massachusetts Bay, the issue of getting to the meetinghouse became a familiar point of contention throughout the colony and likewise within the settlement of Newton. Early in the eighteenth century, surveyors determined that the center of the town actually lay near what is now the intersection of Homer and Centre streets (Fleishman 1999, 7). The citizens agreed to move their meetinghouse to this site thus consolidating another cluster of farmers into a village called Newton Centre. When settlers living due west of the first Jackson enclave complained about the difficulty of getting to the new meetinghouse they got permission in 1781 to build a Second Parish meetinghouse in West Newton.

Settlements at Upper and Lower Falls displayed a similar tendency to coalesce into villages but not on account of meetinghouse controversy. Their choice locations along the waterfalls of the Charles River attracted mills and factories from the earliest days of settlement (Fleishman 1999, 7). Although manufacturing grew up in other sections of Newton, Upper and Lower Falls long retained their discrete character as the centers of industrial activity in the town.

The habit of viewing a certain district as a village within the main settlement continued until Newton filled out its borders. By the time the Stanleys arrived, it had almost reached its present day configuration. In 1890, the City of Newton consisted of twelve distinctive, often contentious but not politically autonomous, villages. Because it was commonplace in Massachusetts for new towns to break away from a parent community, particularly following a meetinghouse controversy, the fact that in Newton so many village offspring remained politically bound together was unusual. Newtonites—as they still do today—tended to think of the various villages as significantly different in character and style from each other. Augusta Stanley recorded in her diary one of her own adventures in crossing village lines. Having ventured from her home in Newton Corner to a bridge party in West Newton, she commented with the surprise of an explorer that "the ladies of West Newton are exceedingly bright and interesting and very good bridge players" (Jan 11, 1907).

The change from simple New England town to fashionable suburb started imperceptibly in Newton. By the 1830s, some Americans had the leisure and the means to vacation (Stilgoe 1988, 24-25). During the summertime, wealthy urbanites began taking to the green borderlands beyond the big cities. Newton with its salubrious air, rolling terrain, and tree-lined roadways, proved to be a popular destination for Bostonians. While these summer visitors did not seriously disrupt the quality of life or significantly affect the growth of the town, they did "precondition" the move to the suburbs that would soon be underway (Stilgoe, 57).

In April of 1834, the Boston and Worcester opened a rail line from Boston through several of Newton's north side villages—Newton Corner, West Newton, and flag stops at Newtonville and Auburndale. The advent of rail service now made it possible to live year-round in Newton even if one did business in Boston. Fifty-five years after the railroad first came to town ninety farms still lay within the city limits (Rowe 1930, 271). But the character of the place was so changed that the sight of "the beauty and fashion of Newton" mingling on the street with "its plain old folks and broad-shouldered farmers" was picturesque enough to warrant mention in *King's Handbook of Newton* (Sweetser 1889, 44).

This transformation might not have occurred, however, if the management of the Boston and Worcester had gotten its way. The railroad wanted the tracks to pass through Watertown or Waltham, but vigorous opposition in those towns, coupled with the lobbying efforts of William Jackson, candle manufacturer and scion of one of the town's founding families, brought the right-of-way through Newton instead. Several times a day, two little steam engines, each pulling a short train of cars that looked like stagecoaches set high on iron wheels, made the roundtrip from Boston. Brakemen perching in coach boxes on each car slowed the train by pressing their feet on levers attached to the wheels (Sweetser 1889, 43). With no mechanical mishaps, the running time to Boston was about thirty-nine minutes, and a one-way ticket cost a little less than a penny a minute—$.37 1/2.

Although early commuter service was both poky and infrequent, it spurred real growth of the town. From the first federal census in 1790 (1,320) until 1830 (2,377), Newton's population had increased by barely over a thousand residents. In a single decade from 1840 (just six

Before the tracks of the Boston and Worcester RR were depressed, man, locomotive, and beast traverse Newton Corner in parallel lines.

Newton History Museum, Newton, Mass.

years after the railway went through) to 1850, the population jumped by more than nineteen hundred. If that growth rate seemed impressive, it was merely the beginning of expansion. In 1850 the total population of Newton was still only 5,258.

The oldest and most populous village of Newton, where John Jackson laid out his first farm, was known familiarly as Angiers Corner and renamed Newton Corner by the railroad. As the first stop on the line, this village experienced the town's first subdivision. William Jackson apparently had anticipated the relationship between rail service and real estate growth when he lobbied the Boston and Worcester. In the 1840s, he divided some of his family land into residential lots and put them up for sale. Jackson's subdivision, however, differed dramatically from the monotonous tract housing familiar in the twentieth century. The newly created lots were laid out around a tree-lined oval park (Newton Department of Planning 1978, 5). Small estates in the Greek Revival and, to a lesser extent, Gothic Revival style began to be built on some of the oldest farmland in Newton.

During the presidencies of Buchanan and Lincoln, more trains ran between Boston and Newton and at greater speed. As a result, Newtonville, West Newton, even Auburndale, which had been a sparsely populated farming village before the railroad came through, were growing apace. With each round of expansion, more large farms were parceled into residential lots. Now Italianate villas and mansard-roofed houses began to supplant Greek Revival as the architecture of choice.

In 1852, textile machinery magnate Otis Pettee had persuaded the Charles River Branch Railroad to extend its tracks from Brookline to his mill works in Upper Falls, thus bringing rail service to villages on the south side of Newton (Fleishman 1999, 34). Trains stopped in Chestnut Hill, Newton Highlands, and Newton Centre before reaching Upper Falls. But the railroad laid only one set of tracks on this route, and the resultant service was so slow that it may actually have retarded the growth of these villages (Rowe 1930, 165). Between 1859 and 1869, the line was used primarily to haul gravel from Needham through Newton to Boston for the mammoth project of filling in the Back Bay. Whenever a gravel train labored past, all other rail traffic had to be sidelined.

With the Back Bay project completed, commuter service improved during the 1870s, and the south side villages at last began to grow. In fact, the town's surging population in all quarters caused the most dramatic political upheaval since official separation from Cambridge. Simple New England town government no longer seemed sufficient to handle Newton's rapid growth. In the spring of 1873, at the annual town meeting, the citizens of Newton voted to have their selectmen apply for a city charter from the state legislature. The following year, ward lines were drawn and the first mayor, James F. C. Hyde, elected. A mayor, board of aldermen, common council, and a school committee now governed Newton.

The city was not united geographically, however, until the mid-1880s when the Boston and Albany Railroad (the former Boston and

Worcester) laid track from Newton Highlands to the Riverside station on its main line from Boston. This much anticipated circuit at last connected the south and north sides of the city by rail. About the same time as it was laying new track, the Boston and Albany began investing its surplus profits in an ambitious station beautification program (Ochsner 1988, 109). The railroad hired esteemed architect, H.H. Richardson, to design a series of depots along its route from Boston. Richardson worked in concert with landscape architect, Frederick Law Olmsted, to create stations that were aesthetically integrated in every detail with their surroundings. The results of their collaboration were successful enough to be dubbed "railroad gardening" by the periodical *Suburban Life*. Richardson died before drafting plans for all the depots, but the firm of Shepley, Rutan, and Coolidge carried out the spirit of his designs. Today, only the Wellesley Hills station, on what is now Boston's commuter rail line, remains with both Richardson's fieldstone building and Olmsted's landscape design in tact.

Although it was the preeminent cause of Newton's growth, the railroad's presence in the life of the city had decided drawbacks. Station beautification could not alter the fact that coal-powered steam trains were extremely sooty and quite noisy. For another, the first tracks through the city were laid at street level, traversing the same space as all other animal and human traffic. This meant that a pedestrian in Newton Corner might literally walk alongside a locomotive as he strode through the business district of his village. Not until the end of the 1890s did this potential for mayhem, urged along by persistent pressure from village improvement societies, result in the huge citywide undertaking of lowering the tracks and eliminating grade crossings.

For a time, convenience seemed to outweigh a little dirt and evident danger. To most people, increased rail service throughout the city made living in the Newtons an ever more attractive proposition. Although urban in its form of government, Newton bore scant resemblance to bigger, older cities in the Commonwealth. Indeed, with its increasingly middle-class population, its distinct preference for single family dwellings,

F.O. Stanley built his first Newton home at 165 Hunnewell Avenue in 1894—quite a step up from Maple Street. *Stanley Museum Archives.*

and its dedication to green spaces, Newton was emerging as one of the premiere suburbs of Boston.

By 1890, Newton embarked on the decade of its greatest expansion. Wealthy newcomers reflected the flamboyant optimism of economic boom times in the extravagance of the houses they built. Their architectural preferences ran to the dramatic turreted facades of the Queen Anne style rather than the elegant symmetry of the mansard. Until the end of the century, the Victorian "painted lady" dominated the architectural scene in the elite neighborhoods of the city. The Stanleys, however, who eschewed excess of any kind, built their Newton residences in the newly emerging colonial revival style. The symmetry of its boxy lines and classical architectural gesture—columns, balustrades, cornices, Palladian windows—appealed to their engineers' sense of order and their New Englanders' sense of restraint.

Despite its growing reputation as an elite suburb, all the neighborhoods of Newton did not consist of Victorian mansions

(Fleishman 1999, 51). Vestiges of an earlier history were evident throughout the city. The Oak Hill section remained predominantly rural, and even fashionable West Newton still contained one of the largest dairy farms in the state. Mills and factories persisted in the villages of Upper and Lower Falls and Nonantum. *King's Handbook,* written in 1889 to foster the image of Newton as a suburban haven, merely hints at the presence of a working class population. The author describes only the most respectably modest sections of the city like the area west of Galen Street in Newton Corner where "acres of neat cottages" stood each with "a bit of green lawn" (Sweetzer 1889, 58). While conditions for the working classes were markedly better in Newton than in Boston, not all laborers were tucked away in comfortable cottages or owned their own dwellings. Some families lived in cramped quarters over storefronts or in tenements. Many single male workers lived in rooming houses and single women commonly lived in the homes of the people who employed them as domestics. In Nonantum, where successive waves of the city's immigrants settled, small end-gabled cottages, built so closely together that grass could scarcely grow between them, represented the predominant architectural style. Streets in Nonantum often lacked the amenities of lighting and paving which typified the wealthier precincts of Newton (Fleishman 1999, 55).

In 1889, the largest ethnic contingent of workers in Newton came from Ireland (2,900). This was more than twice the number of the next most populous group from French Canada (1275). There were only five Italians counted in the city census of 1889, but by the end of the century their numbers would grow substantially. Though largely unnoticed by the city as a whole, there was a well-established Black neighborhood in West Newton with the Myrtle Baptist Church on Curve Street at its heart. The congregation was organized in 1874, but Blacks had lived in Newton since early colonial days. Free black men migrated from Charlestown and the West End of Boston, perhaps drawn by the powerful abolitionist sentiment in West Newton. After the Civil War, a handful of Jews, who made their living primarily as tailors and junk dealers, moved into Nonantum—the only village where the city would license a scrap metal yard. These few families tended to live separately but unmolested—except for taunts endured by their children at school (Gordon 1959, 25). By 1912, their congregation was large enough to warrant the building of a small synagogue, Agudus Achim, still standing at 168 Adams Street in Nonantum.

Not surprisingly, these less affluent, alien segments of Newton's population did not warrant mention by the city's boosters, bent on drawing the wealthy middle class to town. Local promoters preferred to extol the considerable natural beauty of the place. Newton's location among rolling hills had long been considered choice, but by the late nineteenth century in the pages of local prose, the hills clearly numbered seven, and the city was likened, however incongruously, to Rome. *King's Handbook* unabashedly declared Hunnewell Hill "the Capitoline...of this little Rome" and deemed Washington Street, the main east-west route through town, "an imperial highway" (Sweetser 1889, 50). Massachusetts Governor Alexander H. Rice opined that Newton would become "the

The F.E. Stanleys moved to their third and most elegant Newton address, 638 Centre Street, in 1896.

Stanley Museum Archives.

Newton High School in Raymond Stanley's day. During the 1960s the classical structure was razed to make way for Newton North—one of two high schools now serving the city.

Newton History Museum, Newton, Mass.

Belgravia of Boston." Even a few disinterested observers took time to admire the view. General George Ihrie of Grant's Western War Staff fairly gushed over the charms of the growing city: "This nest of Newtons is one of the most beautiful spots on this earth, and reminds one of the suburbs of Paris"(Sweetser 1889, 32). One Dr. Elias Nason summed up the attributes that made Newton the quintessential suburb in terms that met no challenge from satisfied residents: "The society is intelligent, refined, and elevated; the civic advantages are numerous; the railroad facilities are excellent; the climate is healthful; and happy is the man who owns a homestead in this progressive town" (Sweetser 1889, 32).

Hyperbole notwithstanding, Newton by 1890 was an impressive city. With a population of 23,000, it ranked as the eighteenth largest in Massachusetts. But its expansion—in contrast to the chaotic growth of older, manufacturing centers—appeared to be proceeding in orderly fashion. Here indeed was the rational community of middle-class dreams. Newton had already solved or had plans to resolve many of the problems that made urban living so intolerable: it had a good network of roads, widened, re-graded and macadamized. Gas lamps lit most of the city's streets. Horse cars with their "tintinnabulating bells" ran on the major thoroughfares, though after 1890, the electric trolley would replace the equine transports. A police force consisting of three officers and twenty patrolmen kept the city under control. Except for the occasional drunkard, the suburban peace remained relatively undisturbed. By urban standards of the day, Newton had an enviable fire department with a roster of seventy men, seventeen horses, three steam fire engine companies, four hose companies, one truck company, 606 hydrants, and an extensive alarm system.

In the 1870s, with the permission of the state legislature, Newton began pumping its water supply from the Charles River. Filtration beds, dug along the Charles ten feet below the depth of the river, collected water as it percolated through the bottom and sides of the basins. This water was then pumped to a reservoir and into homes through miles of water mains laid beneath the streets. The Newton Statistical Yearbook for 1891 noted that there was the "usual number" of typhoid fever cases in the city. It would be interesting to know what comparative medical statistics are implied in the phrase "usual number." The fact that Newton used filtered river water for its city supply meant that the incidence of typhoid, a waterborne contagious disease, had to have been lower there than in communities relying on wells or untreated surface water. In 1889 the annual death rate from all causes in Newton was below 14 in 1000.

By 1891, Newton planned to make cesspool cleaning a thing of the past by hooking up to the metropolitan sewerage system then under construction. The city budget provided $5,000 annually for the removal of ashes from residents' coal burning furnaces and in 1891 purchased a central dumping site for the disposal of kitchen offal. In an early, if dubious, recycling effort, farmers were to cart away the swill to feed their hogs.

By the time the Stanleys came to town, Newton already spent more money on the public education of each child ($28.71) than any other city or town in Massachusetts. The flagship of the system was the high school in Newtonville, which would prove good enough for Frank and

Augusta's only son Raymond to attend for his Harvard preparation. But educational opportunities for Newtonites were not confined to the schools. All city residents had access to a 20,000 volume free library in Newton Corner. The quality of this collection moved local historian Henry K. Rowe to rave in his 1930 *Tercentenary History of Newton*: "Who would have dreamed of such a collection outside of the British Museum." Two weekly newspapers, the *Journal* and the *Graphic* kept tabs on the city's business from their offices in Newton Corner. Throughout the city there were numerous clubs and organizations dedicated to self-improvement and home study. A Newtonite, man or woman, could find a club or committee to address any political or social issue of the day. Advocates of Civil Service reform had a local chapter. Men who supported President Cleveland's plan for retrenching the nation's highly protective tariff system met as the Tariff Reform Club. The Equal Suffrage League was organized in West Newton during the 1880s. Sentiment stirred by the abolition of slavery in the Civil War led to concern for the status of the Native American and formation of the Indian Rights Association. The chronic issue of national sobriety found a voice for temperance in the Prohibition League of Newton.

On the lighter side, there were plenty of activities to entertain the newly leisured classes of Newton. The introduction of the safety bicycle in the late 1880s extended the cyling craze to women as well as men, although an oft-told bit of Stanley lore maintains that Augusta Stanley never could learn to ride a bicycle. This supposedly prompted Frank to build her a motorcarriage. For all Newtonites, the Charles River at Auburndale became a favorite spot for outdoor pursuits like swimming, canoeing and ice-skating. The Charles was the scene of summer band concerts, boat races, and lantern-lit flotillas. Cricket and baseball clubs played in the open fields. Courting couples picnicked on the riverbanks nearby.

By 1890, a dozen trains served Newton daily. The morning train from Boston brought the mail, which postmen delivered to the doors of all residences with street numbers. At night the outbound trains brought the husbands home, often bearing parcels purchased by their wives on

Boston shopping excursions. While the women shopped in the city for fashions and furnishings, or ordered groceries from S.S. Pierce for special occasions, they had all the daily necessities within easy reach of their back doors. Icemen and milkmen made deliveries to Newton kitchens. The business district of Newton Corner, for example, offered fish and meat markets, provisioners, and green grocers (though Augusta Stanley thought their prices too high and when economizing preferred to buy her groceries in Waltham).

Although the Stanley brothers did not commute to Boston for business, they must have appreciated rail service in such close proximity to their factory. Like the other "quality folk" (to borrow Augusta Stanley's term) who had settled in Newton, the Stanleys were seeking the numerous genteel amenities that the city offered. In return, the brothers and their wives, as their forebears in Kingfield had done, delved into the civic life of their adopted community. But unlike earlier Stanley generations burdened with financial failures, the twins accompanied their rise in community stature with steady profits throughout the 1890s.

CHAPTER 4

Business is Simply Immense

IN MARCH OF 1890, the brothers began building the new headquarters for the Stanley Dry Plate Company off Galen Street in neighboring Watertown. They had bought a lot near the Charles River in an industrial zone Newtonites then called "the Watertown District of Newton." This designation referred to a geographical not jurisdictional anomaly dating back to colonial times and the granting of fishing rights. In 1705, committees from both towns confirmed that Watertown retain certain weir lands on the south side of the Charles (*Watertown Records* 1909, 160). While the river served as a boundary between the towns with most of Watertown north of it, this 150 acre tract between Newton Corner and the Watertown Bridge officially became part of Watertown (Rowe 1930, 43). Since this sector was contiguous with Newton, it was possible for the Stanleys to use Newton city water at their dry plate factory and convenient to utilize Newton rail and postal services. The first U.S. post office in the city was located about three-eighths of a mile from the Charles on the northeastern side and about three hundred yards north of the Boston and Albany Station in Newton Corner (National Archives Microfilm 1986). Thus, while the Stanley factory stood in Watertown, their company letterhead and F.O.B. read Newton.

At the time the new factory opened, the Stanleys were not quite in the top echelon of dry plate producers. Eastman of Rochester, Corbutt of Philadelphia (Keystone Dry Plate), and Cramer of St. Louis had emerged as the most powerful and influential manufacturers by the end of the 1880s. The Stanleys, however, were right behind the leaders and doing well enough to expand their operation. As earlier noted, just before moving their headquarters to Massachusetts, they opened a Montreal branch to accommodate photographer William Notman.

Interestingly, George Eastman did not attain front rank because of the superiority of his dry plate. In fact there was a period in the early 1890s when his company had a reputation for inferior emulsion (Jenkins 1975, 154). What distinguished Eastman from all the other dry plate manufacturers—what kept him at the forefront of the photographic industry—was the effort he made early on to innovate and diversify production. He did not settle for manufacturing dry plate alone but began to expand vertically, producing bromide paper and photographic enlargers in the early 1880s. By the end of the decade he had hit on the invention that ensured his future mastery of the photographic industry.

In the late 1880s Eastman brought out the first of his "Kodak" hand-held cameras. He understood that if business were to continue to grow, photography had to lie within reach of the most casual amateur. He even concocted the name "Kodak" from a combination of vowels and consonants that he believed pronounceable in any language (Brayer 1996, 61-63). Eastman in fact succeeded in designing a "little roll holder breast camera" that anyone could use—for the price of $25. Totally portable, needing no tripod, each camera came equipped with a 100-exposure roll of film. There was no viewfinder and the whole contraption had to be sent back to the factory to develop the pictures (Brayer 1996, 61-63). Primitive though it was, this little box camera foretold the eventual demise of the dry plate industry.

During the 1890s, however, glass still dominated the market, and Eastman, despite his new interests, kept actively involved with dry plate manufacture. Leading by example, Eastman, Seed and Cramer forced most of the major companies into the highly charged quest to find a European chemist who could produce a winning emulsion formula. The Stanleys, true to their stubborn form, persevered on their own.

Just three years after the twins relocated to Massachusetts and ten years after the first economic downturn wiped out many struggling companies, the dry plate industry faced another financial crisis in the Panic of 1893. During the previous recession, the biggest dry plate manufacturers and dealers had formed a consortium in order to prevent a price war. When the economy began to recover and profits again rose, the association disbanded. Now in the face of another financial upheaval, the Dry Plate Manufacturing Association reconvened in January 1894, first in St. Louis then two weeks later in New York City. The Stanleys came to those meetings with a production problem unbeknownst to their competitors and a personal agenda tucked in their vest pockets.

The brothers had discovered that the Newton water they were using to wash the glass for their dry plate was deficient in calcium carbonate. The absence of lime in the water meant that the dry plate lost its shelf life. According to Emma Walker, Augusta Stanley's sister who worked as the bookkeeper and cashier at the Stanley factory, by 1893 the Stanleys had between $30,000 and $50,000 worth of bad plate on the market. In a letter dated May 24, 1936, Walker went on to explain to her nephew Raymond Stanley that the brothers then sank a three-hundred-foot artesian well through solid ledge in order to rectify the water problem (Walker 1987, 10). To set right their bad inventory they simply rolled up their sleeves. As Emma Walker explained, "(T)hey knew they would have to make them good to the dealers or face a good many lawsuits."

While the Stanleys came to the union meeting with concern about their manufacturing integrity, they, like the rest of the association members, had to grapple with the problem of falling sales in the face of recession. As F.O. explains in his retrospective history of the Stanley Dry Plate Company, "The trade of photographic stock was so reduced that dealer profits, always small, owing to competition, would barely pay expenses." A majority of the association members favored raising prices to offset declining sales. The Stanleys, in the minority, objected to the plan with the argument that fixed prices along with the decreasing cost of materials would bring many new competitors into the field (Jenkins

1975, 220-221). That night while others went out to the theater, the brothers repaired to their hotel room to study "the Dry Plate problem." One can fairly imagine them posting a 'Do Not Disturb—Crafty Yankees at Work' sign on their door. The next morning, in F.O.'s telling, Frank Stanley electrified the assembly when he announced that Stanley Dry Plate would agree to raise prices. But, F.O. adds, no one knew "our real plan" (F.O. Stanley [1936] 1987, 9).

At this juncture, the story told by F.O. does not quite jibe with Emma Walker's account. By Freel's telling, the brothers came back to Massachusetts and "went to work" obtaining the names and addresses of every photographer in America. They then notified all photographic dealers that they intended to quit the association to sell directly to customers—at prices well below the schedule set by the association.

Emma Walker, however, describes a slightly different sequence of events. As the person overseeing the daily business minutia of the company, her account seems entirely credible. She claims that the Stanleys first got to work making good on the ruined dry plate and parted with the association only after they had made complete restitution. By her account, they took almost no new orders during the summer of 1894 in order to replace the damaged stock. Walker concluded that the Stanleys needed to remain in the union until they made things right: "It was better for the Stanley Co. to stay in the union and replace dry plates with dry plates than to get out of the union and pay cash for the bad plates...(T)hey probably did not lose more than $10,000—and kept their business integrity" (Walker 1987, 10).

By Walker's timetable, the Stanleys had enough confidence in the quality of their dry plate during the fall of 1894 to send out circulars to every American photographer and to establish cut-price dealers "in all the large cities north of Washington." According to Walker, the Stanleys did a lot of cut-price business. To her bewilderment F.O. never admitted to anything except direct sales to photographers: "If they had been dependent on sales direct to consumers they would have had a small percentage of their trade...It was perfectly legitimate to establish cut-price dealers and I don't see why F.O.S. always maintains that the Stanley Co. sold exclusively to photographers when they left the union" (Walker 1987, 10).

Apparently, Emma Walker never fully grasped the brothers' stubborn conviction that their products sold themselves without resorting to deals or gimmicks. Freel's selective memory about the company's cut-price business may derive from that cherished belief. It is a bit more puzzling why F.O. neglects to mention that the brothers used their membership in the association to help buy time for what could have been a manufacturing disaster. Surely, this part of their strategy was as shrewd as their eventual departure from the union.

Whatever the actual order of events, the Stanleys' move to quit the union proved eventually to be a marketing coup. In the beginning, however, when all the photographic dealers boycotted the Stanley product, prospects looked bleak for the company's survival. In F.O.'s terms, "the dealers were certain they had licked us" (F.O. Stanley [1936] 1987, 9). But the Stanleys managed to hang on until their direct marketing campaign began to pay off. For without the agency of a middleman, they could offer photographers cash transactions "at the lowest prices ever," forty percent off list price, according to F.E. (*Stanley Museum Quarterly* 1991, 12). The company even paid shipping costs on orders of a caseload or more.

Business boomed for the Stanleys without the jobbing houses (Jenkins 1975, 220-221). After months of a standoff, it was the dealers who conceded defeat. Francis Hendricks & Co. of Syracuse broke the boycott with a directive to the Stanleys, "fill our order." In quick succession, the other big houses fell into line. In a letter dated May 16, 1896, Frank Stanley wrote to his wife Gustie, who was traveling in Europe, "business is simply immense." "We shall carry over to June orders for more than 1200 cases of plate that we could not fill and we shall sell over $50,000 list"(Stanley Museum Archives).

Soon the Stanley Dry Plate Company was grossing $60,000 monthly; before the union pullout the figure ran closer to $20,000. By mid decade, Stanley had moved up to challenge the current leading dry plate manufacturers: Seed, Cramer, and Hammer. Close on the heels of third place, Stanley forced Hammer to offer a deeper discount to dealers in order to stay ahead. Eastman, again struggling with emulsion problems, had fallen back in the dry plate market (Jenkins 1975, 225).

It is ironic that the Stanley brothers, who were not known in photographic circles for their creative business practices, pulled off the marketing coup of the decade. The twins who scorned mass advertising and eschewed diversification toppled the price-fixing consortium of 1894. The Stanleys' refusal to stay in the Dry Plate Association and adhere to the price-fixing agreement is the largest single reason that the consortium collapsed (Jenkins 1975, 220). One could argue that it was Yankee mulishness not imagination that led to their triumph. If true, the twins' contrary instincts served them well in this instance. By flouting conventional wisdom, the Stanleys proved conclusively that jobbing houses had outlived their usefulness Their downeast stubbornness clearly won them a business advantage.

CHAPTER 5

Our Price is as Stated

BY THE TURN OF THE CENTURY, the Stanleys were making the largest profits of their career, grossing $100,000 a month with a company ranked third among the leading dry plate corporations (Jenkins 1975, 225).

Interestingly, despite their impressive ranking, the Stanleys controlled only seventeen percent of the American dry plate market. The first-ranked Seed Corporation monopolized fully fifty percent of dry plate sales—evidence that during the years since the failed consortium, the dry plate industry had become increasingly oligopolistic. Throughout the late 1890s, those manufacturers still in business also had to contend with steadily dropping prices (Jenkins 1975, 225).

George Eastman, whose company commanded only six percent of the dry plate market, now proceeded as if this sector of the photographic industry was in its death throes (Brayer 1995, 202-3). He made no effort to increase market share by improving his emulsion formula as he had done a few years earlier. Although he was unwilling to commit to dry plate research, Eastman did not ignore this sector altogether. While there was profit still to be made, he was in the competition. Starting in 1899, he promoted the idea that the four leading dry plate manufacturers (Seed, Cramer, Hammer, and Stanley) form a combine with him as sole agent (Brayer 1995, 203). At the Paris Exhibition of 1900, he convinced Gustav Cramer of the soundness of this plan, and, when the cost of Belgian glass went up the following year, interest in Eastman's idea widened. By this time, however, Eastman had begun to talk not of combines but buyouts.

With the example of the defunct price consortium before him, it is hard to believe that Eastman seriously considered a voluntary combination among dry plate manufacturers. More likely he decided early on to buy out his competitors. He certainly wasted little time in putting together a holding company for that purpose. Even before the new parent-corporation was formally launched, Eastman and his vice-president Charles Abbott came to Boston in August 1901, to find out whether "that great pair of Yankees," the Stanley brothers, could be induced to sell.

According to Abbott the twins were eager to do business (Brayer 1996, 203), but negotiations were soon hopelessly stalled. In the meantime, Eastman Kodak of New Jersey became fully operational. During the spring and summer of 1902, with the Stanley transaction in limbo, Eastman successfully acquired both the number one ranked Seed Corporation and Standard Dry Plate of Lewiston, Maine, acquiring with the latter the expertise of acclaimed chemist Milton Punnett.

Eastman approached all potential buyout companies with the same basic formula: he offered to pay five times the average of three years' profit (this recognized the company as a going concern) plus the value of live assets at inventory cost. In his buyout offer to the Stanleys, however, Eastman altered that formula by proposing to calculate the figure for the "good will of the company" solely on the net profitability of 1901. Because he felt that the Stanleys had overvalued their plant assets and underestimated the cost of superintending production, Eastman requested that before multiplying the profit figure by five, the brothers subtract six percent of the assets and a reasonable sum for superintendence.

F.O. Stanley, as treasurer of the corporation, responded to Eastman's directive largely by ignoring it. Without addressing the issue of the cost of superintendence or defending the asking price of the company assets,

F.O. put forth a counter offer in his letter of June 2, 1902. The Stanleys preferred to calculate their profit figures on an average of three years because 1901—due to difficulty obtaining glass—was less lucrative than either the year preceding or the current year. Valuing their assets (including the Montreal plant) at $200,000, the Stanleys set a price of $650,000 provided that the business would be transferred to Kodak on or before July 1 (*Stanley Museum Quarterly* 1994, 12-13).

Eastman replied with a request to audit the Stanley Company. On June 5, 1902, F.O. penned a somewhat starchy reply: "The profits and assets will not vary materially from the amount stated in our letter. Our price is as stated." On the ninth, F.O. pressed Eastman for a commitment to buy (*Stanley Museum Quarterly* 1994, 13).

Eastman must have issued an expression of good will and intent because F.O. Stanley subsequently agreed in a letter dated June 12, 1902, "to let the matter rest just as it is to suit your convenience" (*Stanley Museum Quarterly* 1994, 13). Eastman then dispatched his auditing team to Watertown. W. Stubbs from Price, Waterhouse was the man on the scene and sent a preliminary assessment back to Rochester on July 31:

> ...I have never seen a stock of goods in so bad a condition, also so hard to get at. They have no inventory book and do not seem to care whether we get one or not. They have in the neighborhood of 15,000 cases of glass all sizes and brands lumped together in all parts of their works. I find that the first floor which is piled with glass is braced from below with the glass that is in the cellar (George Eastman House Archives).

Stubbs goes on to explain that he is reluctant to move any of the glass-containing cases in the basement due to "the light construction of the building," and concludes that at best he can only estimate the value of the stock in the basement.

After examining the Montreal plant as well, the Eastman audit team calculated that the Stanleys had overstated their assets by $7500 and their profits by $10,000-15,000. In a letter dated December 1, 1903, to

F.E. Stanley, Eastman explained that he had expected "to obtain some offset on these shortages," but when none was forthcoming, he withdrew the offer to buy (George Eastman House Archives).

Eastman and the Stanleys exchanged no further buyout proposals until November of 1903—almost a year and a half later. Eastman wrote to George Abbott on October 10, 1903, that he intended to reopen negotiations with the Stanleys after learning via the grapevine that they were sorry they had not sold out to him in 1902 (George Eastman House Archives). The next month while in Boston, Eastman did indeed test the buyout waters and sent Abbott a letter dated November 27, 1903, with a round-by-round account of his meeting with F.E. Stanley (George Eastman House Archives). Over lunch, F.E. confided that his brother was in failing health and that he would be out of the business for at least a year. At this Eastman sensed a bargaining advantage, but then Stanley countered with another bit of unexpected news. In Eastman's words, Frank let slip "in the coolest, offhand manner imaginable," that the brothers had hired H.C. Ross away from Standard Dry Plate—one of the firms Eastman had recently acquired. F.E. assured his old friend George that Ross knew Milton Punnett's emulsion formula and that Stanley intended to put out a plate identical to Standard's. F.E.'s casual pronouncement was enough to prick Eastman's competitive instincts, and now he was eager to deal.

However much Eastman wanted the Stanley name, the audit from 1902 had convinced him that the Stanley business was in great disarray. When negotiations officially resumed, therefore, Eastman withdrew the offer to buy the Stanley Dry Plate real estate. This was apparently acceptable to F.E. who (as will soon be evident) had new plans for the Watertown plant. In a letter dated November 28, 1903, Stanley reiterated his original offer, minus the Newton plant. "...(W)e will sell our dry plate business on the same terms and for the same price which we made to you in June 1902... and will allow you for said real estate, etc., the sum of Eighty Five Thousand Dollars." He further asserted that "the sales for 1903 are about the same as they were in 1902...therefore our business is worth as much to us now as it was in June of 1902." While F.E. agreed to permit Eastman auditors to verify current sales figures, he adamantly refused a repetition of the previous examination: "...we will allow...no such prolonged and exhausted (sic) investigation which you made to ascertain what our profits have been" (*Stanley Museum Quarterly* 1994, 17-18).

Clearly smarting from the auditor's unflattering portrait of the Stanley business, F.E. continued to complain about the audit Eastman had run in a subsequent letter dated December 3, 1903. "I do not know how you could have determined what our assets were," he wrote, "for you did not take the trouble to determine how much stock we had." He dismissed the apparent discrepancies between profit figures declared by the Stanleys and those calculated by the auditing team with this disclaimer. As a condition for the sale of the company, "we were not willing to allow any reduction in what we called profits on account of cost of management." It is clear by the end of the letter, however, that these issues had been dropped from the current negotiations. After a lengthy berating, F.E. brings the letter to a hastily conciliatory close: "However, this is unimportant now since we have come to an agreement at the present time" (*Stanley Museum Quarterly* 1994, 19).

One unresolved problem, however, still held up the final sale: what to do about Stanley employees who were privy to the firm's emulsion formula. Eastman's main concern was that these men agree not to work in the dry plate business for ten years unless employed by Kodak. For their part the Stanleys wanted assurance that Eastman would take good financial care of H.C. Ross, and their two nephews, Newton and Carlton (both sons of Isaac Stanley). Newton had been the emulsion maker for the firm before he was seriously injured in an automobile accident, and his younger brother Carlton managed the Montreal branch of the business. In a letter dated December 28, 1903, F.E. agreed to have these three employees sign a contract abstaining from the dry plate business for ten years provided Eastman employ them "for five years at the same salary they are now receiving"(*Stanley Museum Quarterly* 1994, 20).

With that arrangement the last impediment to sale was eliminated. In January 1904—fully three and one-half years after the first round of negotiations—Eastman Kodak purchased Stanley Dry Plate. Why the deal took so long to consummate points directly to a major difference between Eastman and the Stanleys on business practices. The Stanleys never credited or perhaps comprehended Eastman's contention that the "cost of management" had to be deducted from profits. As the owners of the company, the Stanleys considered themselves the management and therefore did not see the need to account precisely for that expense. Reese Jenkins concluded in his analysis of the buyout that Eastman at first balked at the purchase of the Stanley Corporation because that company represented "considerable expense and little technical advantage." According to Jenkins, "the unsophisticated character of the company's production and accounting systems did little to entice" Eastman to buy (Jenkins 1975, 229). But for all the seeming backwardness of the Stanley operation, the company continued to hold its strong position in the market, and Eastman felt compelled to buy out the Stanleys.

The final selling price for the company was closer to $540,691 (the Stanley figure quoted in a letter dated December 3, 1903) than $650,000 (the original asking price). Though the deal did not fall to the Stanleys' advantage, eleven years after the sale Frank Stanley remembered things differently. Testifying on May 25, 1914, before the Senate Anti-Trust Hearings into the business practices of the Eastman Kodak Corporation, F.E. said he "did not recall the total price" for the sale of his business. But he was sure that it cost Eastman "more than it would have a year before because the last year our profits had been greater"(*Stanley Museum Newsletter* 1991, 13).

After the Stanley deal was closed, Eastman made no more dry plate buyouts. Hammer and Cramer, who had initially been so interested in Eastman's plan, continued as independent entities holding equal market shares with Kodak (Jenkins 1975, 230-33). The old Stanley Dry Plate Company was dissolved April 13, 1905, but the Eastman Corporation continued to manufacture plates under the Stanley brand name at its Rochester plant until 1921. Then, as a result of the government's anti-trust suit against Kodak, Eastman divested itself of the dry plate, plate cameras, and platinum paper divisions—all of which, according to Eastman biographer Elizabeth Brayer, "were in the dying sectors of photography in 1920."

It is not clear whether the Stanleys perceived or cared about the waning nature of the dry plate industry back in 1901 when they first entertained the idea of selling to Eastman. The more likely incentive for them to sell derived simply from the fact that the brothers had lost interest in the photographic industry. By the time Eastman and Abbott approached the Stanleys with an offer to buy, the twins were already deeply involved in the new field of automotive technology. Profit from their dry plate business and subsequently from the sale of that business enabled the Stanleys to get started building cars. As F.O. put it: "There was more fun in riding in a machine of our own make than there was in making dry plates..."(F.O. Stanley [1936] 1987, 9). In fact, by the time of the final sale to Eastman, the twins had sold one car business and started another. No wonder their dry plate inventory was in shambles.

George Eastman and Mrs. Pauline Abbott take a back seat to Kodak vice-president Charles Abbott at the wheel of a Locosurrey. Photograph by William Carter.

Stanley Museum Archives.

CHAPTER 6

All Out of Horses

THROUGHOUT THEIR TOE-TO-TOE NEGOTIATIONS with Eastman, the Stanleys remained on friendly terms with the photographic giant. What kept things amicable may well have been Eastman's admiration for the new Stanley product—lightweight, steam-powered automobiles. Even during the most heated period in the sales negotiations when Eastman requested an independent audit of the Stanley dry plate business, their shared enthusiasm for the horseless carriage broke through to diffuse the tension. On June 12, 1902, F.O. Stanley sent this surprisingly jaunty invitation to Eastman: "We hope when your experts come here to look over our business you will come with them as we have something new we wish to show you in the way of steam carriages" (*Stanley Museum Quarterly* 1994, 13).

At this point, Eastman, an ardent car collector, already owned two Locomobiles—a topless, two-seater runabout and a four-passenger surrey with canvas canopy. He also bought a runabout from the Stanleys after they began assembling cars again on their own. Eastman had asked the twins to equip the car with a double-acting brake. In their customary disregard for outside intervention, the Stanleys declined to comply, thus making Eastman mad (Brayer 1996, 233-4). Apparently, they didn't make him angry enough to cancel the order.

In the earliest days of the automobile industry, the horseless carriage was primarily a rich man's toy, but the American men who first experimented with self-powered steam vehicles came from much humbler circumstances. They were, for the most part, New England farm boys and Yankee tinkerers who saw no reason why steam could not power a vehicle over land (White 1991, 13). In 1791, Nathan Reed, a Harvard professor from Warren, Massachusetts, procured the first American patent for a steam-powered land carriage, but Reed never got his plans off paper. John Gore of Brattleboro, Vermont, is said to have built a viable steam vehicle in the 1840s. A close encounter with a team of horses on a rugged country road caused him to crash. One William H. Harrison wrote the editor of the *American Machinist* in the December 15, 1898, issue to verify that he had viewed the pieces of the wreckage. There is no evidence that Gore rebuilt his carriage (McKay 1986 pt. 1, 8).

The first man in the United States to successfully build and sell several types of "self-propelled road vehicles" was probably Sylvester Roper of Roxbury, Massachusetts (Bacon 1984, 1). Roper began experimenting around 1859 and by 1863 had produced his first steam buggy rigged with a coal-powered boiler. The results impressed the *Scientific American* which reported in its March 14, 1863, edition that Roper had "invented and completed...a very successful" steam carriage. In 1868 he also invented a steam-driven bicycle and a self-propelled fire wagon (Bacon 1984, 2, 17). Interestingly, Roper made no attempt to protect his automotive inventions. Since he held patents on at least twenty other inventions, his failure to seek protection on his steam carriages was not due to naivete. He may have believed (as the Stanleys later maintained in court) that steam vehicles were not patentable.

In fact, Roper readily displayed his vehicles at state fairs and permitted eager and ambitious young inventors to comb over them thoroughly. One such inspired youngster, George A. Long of Northfield, Massachusetts, built his own version of the steam car after a trip to the fair. Long managed to build an automobile that was somewhat lighter and smaller than Roper's, thus departing from the concept of the "road locomotive" (Bacon, 36).

Sylvester Roper of Roxbury, Massachusetts, built this steam-powered motorcycle in 1868—just three years after the end of the Civil War.

John H. Bacon Collection, Stanley Museum Archives.

One of the drawbacks to Roper's design was the use of a coal-fired boiler. It proved impractical and undesirable for road use because it was impossible to regulate the steam pressure once the fire was started. The driver either had to run the carriage until it stopped or blow-off steam through a safety valve. To tend the fire, the operator had to carry his own coal supply slung over his shoulder in a four pound sack and then hand shovel chunks into the firebox as needed. Subsequent steam carriages, therefore, had boilers heated either by gasoline or kerosene burners.

Not all New England inventors concentrated their creative energies on steam. In 1834, a Vermont blacksmith named Thomas Davenport built an electric motor, which he later used to power a small car around a circular set of tracks. By the late 1880s, battery-powered electric cars began to appear as road vehicles. Clean, quiet-running and easy to start, electric autos were very much in the competition to dominate the nascent automobile industry in America. Their lack of exhaust emissions made them particularly popular in big cities. Electric cabs became popular enough in New York City to launch the Electric Vehicle Company in 1896.

Still other inventors experimented with a relatively new form of motive power—the internal combustion engine. Germany's Niklaus Otto, Gottlieb Daimler, and Karl Benz led in the experimental development of this technology during the 1860s and 1870s, but American mechanics were soon competitive. In 1893, the Duryea brothers of Springfield, Massachusetts, built a creditable gasoline-powered car. Their model consisted of a one-cylinder engine mounted on the back of a high-wheeled buggy. Two years later, Frank Duryea drove their invention to victory in the first organized automobile race in America.

On Thanksgiving Day 1895, six "motocycles," as they were dubbed in the press, raced fifty-four miles from Chicago to Evanston, Illinois, and back over snow-rutted roads. Although some eighty people had responded to the initial announcement for the race, only six entrants actually appeared on race day. Two of the cars were electric models; one lost power before the race ended and the other's engine overheated while running through deep snow. Three of the competitors were German-built Benz cars, lightweight three-wheelers powered by a one-cylinder gasoline engine. New York department store magnate, R. H. Macy, who intended to import and sell these automobiles, entered one of the Benz models (Wright, 1996). Macy hoped that a strong showing would promote sales, but his car collided with a hack and had to withdraw from the competition. Only one of the other Benz autos even finished the race—an hour and a half behind the Duryea. The winning "buggyaut," as the Duryea brothers first called their car, averaged around 8 miles per hour, finishing the course in ten hours (including a fifty-five minute pit stop). The driver pocketed $2000 for his effort, and *The Chicago Times-Herald*, whose publisher H.H. Kohlstaat had sponsored the event, declared the following day: "Persons who are inclined…to decry the development of the horseless carriage…will be forced…to recognize it as an admitted mechanical achievement" (Library of Congress).

About the time the Duryeas were at work on their gasoline car in western Massachusetts, George Whitney, great grand nephew of inventor Eli Whitney and nephew of Amos Whitney of Pratt and Whitney engine fame, was working to build a steam car in Boston. Whitney was the automobile inventor most closely linked to the Stanley brothers. He was thirteen years younger than the Stanleys but a few years ahead of them in automobile experimentation. Typical of the new generation of engineers, he had studied at the Massachusetts Institute of Technology School of Mechanic Arts. He began his career working on steam-powered boats but through his association with Sylvester Roper became interested in automobiles (Bacon 1984, 46).

By 1890 Whitney's interest had clearly begun to shift to motor carriages. Whitney told historian John Bacon—though his story cannot be verified by independent sources—that William Mason of Boston commissioned him to design a lightweight two-cylinder steam engine for automotive use. According to Bacon, for a decade, the Mason engine, with a patent only on its valve mechanism, became the engine of preference for American automobile inventors, who either purchased it outright or "tooled up to produce their own engine patterned on (it)." Bacon included the Stanley Brothers with the latter (1984, 47).

Whitney himself built a steam car in 1896. The six-hundred-pound automobile was fitted with rear wheels larger in circumference than the front ones and sported the recently popular pneumatic, or inflatable tires. Gasoline fueled a boiler that was built to withstand 125 pounds of pressure. In June of that year, Whitney attempted to drive from Massachusetts to New York. He successfully negotiated the journey from Boston to Stonington, Connecticut, but a storm foiled his efforts to reach New York under his own power, and he had to ship the vehicle the rest of the way by steam boat (*The Horseless Age* 1896, 5).

In April, 1897, *The Horseless Age,* then a monthly magazine which covered every aspect of the infant auto industry, ran an account of Whitney's driving a second car 130 miles from Boston to Hartford, Connecticut, in ten and one half hours. This model weighed 800 pounds, had rear-wheel drive, and a steam boiler fueled by gasoline. A subsequent story in a Claremont, New Hampshire, newspaper confirmed that Whitney made a test run through that town in hopes of drumming up business. Since a Whitney carriage cost $1500, it generated more excitement than sales among ordinary folk (Bacon 1984 {reprint}, 61-62).

As a very old man in the 1950s, Whitney told John Bacon about the steam car buffs who frequented his shop in those early days (Whitney 1955). According to Whitney's recollection, both Stanley brothers made a few short visits to East Boston between 1895 and 1896, but during those brief encounters Whitney claimed to have given them many pointers on building their first car (Bacon 1984, 62). The Stanleys themselves admitted their innocence on the subject of steam power when they first got interested in automobiles, but Whitney may have exaggerated the importance of his role. As will soon become clear, he had strong personal reasons for believing himself to be the author of the Stanleys' automotive success.

When and how the twins began to build cars are questions that automotive historians have not resolved definitively. Some of the difficulty no doubt arises from the Stanley penchant for telling a good story. Richard and Nancy Fraser, authors of *A History of Maine Built Automobiles 1834-1934*, speculate that the Stanleys may have gotten their first tinkerers' taste for steam cars while still in Lewiston in the late 1880s. It may be there that Frank Stanley had what he later claimed was an inspirational run-in with a steamroller. In an interview printed in the December 1, 1898, issue of the *American Machinist*, he recounts how the sight of a steamroller spurred him to build a steam car. A year earlier he wrote his wife Augusta on July 7, 1897, that he planned to create a vehicle that, unlike a horse, "will not be afraid of a steamroller and will have no bad habits" (Stanley Museum Archives).

More convincing evidence of a Maine incentive comes from Edwin F. Field, an early builder of steam carriages in Lewiston, who recalled for *Motor World* magazine the interest the Stanley twins took in his work during the late 1880s. The brothers stabled their trotting horses on Canal

Street where Field had his shop and, by his telling, frequently dropped in on the inventor (*The Motor World* 1904, 1005). Still, a publication produced by the Stanley Motor Carriage Company itself makes no mention of Field or Lewiston. According to this brief account, possibly written by F.E.'s son-in-law Prescott Warren, Frank is supposed to have caught the automotive bug at the Brockton Fair in October, 1896, after viewing both an electric and a gas car (*Bulb Horn* {1918} 1945, 19).

Even within the Stanley family itself, the issue is open to dispute. In 1930, F.O. wrote an account of the brothers' car business similar to the history he wrote of the Stanley Dry Plate Company. In Freel's version, he claims that the two of them drew up the plans together and built their first automobile between 1896 and 1897. According to his nephew Raymond, however, F.O. took too much credit for the invention of that first Stanley steam car and for subsequent automotive refinements.

With loyalty to his father clearly piqued, Raymond Stanley gathered over his lifetime what he considered to be incontrovertible proof that Frank, not Freel, was the guiding force in the brothers' automobile business, and that his father designed and built that first car by himself. Raymond maintained in correspondence dated October 15, 1942, with F.H. Elliot that patents and letters from 1896-1905 "prove conclusively that credit for the invention...and manufacture of the first Stanley steam car belongs solely to F.E. Stanley" (Stanley Museum Archives). Citing a recurrence of the tuberculosis that had plagued Freel in his earlier days, Raymond concluded that from 1903 on, since F.O. spent most of his time in Colorado for reasons of health, he left "responsibility for the design, improvement, manufacture and sale" of the Stanley Steamer exclusively to F.E.

The patent records that Raymond cites as evidence of his father's technological dominance do show that F.E. Stanley appeared with sole attribution on at least ten automotive inventions while F.O. obtained only two. Yet during that same time frame, F.O. and F.E. were issued thirteen joint automotive patents, which suggests a high degree of collaboration (Gazette, U.S. Patent Office).

Raymond's latter assertion concerning Freel's frail constitution is corroborated in part by Charles Abbott, the Eastman vice-president who first negotiated with the Stanleys in August of 1901. Abbott was convinced that the brothers wanted to sell their dry plate business because Frank was immersed in automobiles and Freelan Oscar was in poor health (Brayer 1996, 203). Entries in Flora Stanley's diary verify that F.O. was in a declining state as early as 1900. On the entry lines for January 31, 1900, Flora wrote: "This year I kept no diary. It has been a year of much sickness and anxiety. Freel had pleurisy in May and was confined to bed for 5 weeks." His condition seemed no better in 1901 with Flora writing entries like this on May 26, "Freel had a very bad night" (Stanley Museum Archives). By February 1903, his condition had deteriorated into a recurrence of tuberculosis. F.O was indeed forced to withdraw temporarily from business with his brother to recuperate in the mountain air of Estes Park, Colorado.

Illness may have kept Freel from being as actively involved in the auto business as his brother, but it does not preclude the likelihood that the Stanleys studied automotive technology together or that they produced their first working models in close cooperation. They jointly attested in a legal brief submitted in 1902 that before building that first Stanley automobile they both were familiar with the large number of steam car patents and texts then available in the United States, including the work of British engineers Walter Hancock and Alexander Gordon. By implication, these were the sources from which the Stanleys taught themselves steam car technology.

F.O.'s participation in the earliest stages of the planning is at least endorsed anecdotally. Back in Lewiston, Edwin Field claimed that Freel stopped by the shop more often than Frank: "Many's the night I have sat in the old shop with Mr. Freeland (sic) and talked to him about a carriage that would go by steam" (*The Motor World* 1904, 1005). George Whitney likewise recalled both twins being in his East Boston shop during the mid-1890s. Harmon Elliott, who may be the most reliable witness of all, remembered that in the days before the brothers marketed their first steamer that "F.O. Stanley spent at least one hour every evening in [the Elliott living room] talking mechanical problems" with his father. Harmon wrote, "I was allowed to listen" (Elliott 1945, 25). One can imagine a young boy, up past his bedtime, struggling to catch every word as the grown-ups wrestled with the problems of a new invention.

While it may never be possible to assign proper attribution for the design of that first Stanley automobile, one thing is certain about its construction. Letters from the brothers to their wives reveal that the car was built for Frank.

During the summer of 1897, Augusta Stanley was on a "grand tour" of Europe. While she was admiring the castles and ruins, F.E. urged her to make more contemporary observations. On June 14, 1897, he wrote her in Paris: "...(K)eep your eyes open to see...motor carriages." He asks her to note the price and type of autos "in common use" among Parisians. He tells her, "I am all out of horses and I shall not own a horse again until I have seen the outcome of the motorcarriage movement." This was literally true. Frank sold his last trotting horse sometime after Augusta sailed for Europe. Several weeks later, in a letter dated July 11, he revealed that the horse's replacement was in the works:

> I wrote you some time ago about motor carriages. Well I am building one. I am making the plans and it will only weigh 350 pounds and will be four inches wider and five inches longer than our buggy. It will cost me about $500 and will be finished the first of September or soon after you get home...It will stand without hitching (and perhaps that is all it will do). (Stanley Museum Archives).

That first Stanley model was assembled in pieces at the Newton Corner buggy shop of P.A. Murray. The latter built the carriage while the American Waltham Manufacturing Company, makers of the "Comet" bicycle, constructed the frame and running gear from the design Frank brought them (McKay 1986 pt. 2, 8). On September 17, 1897, the *Newton Journal,* one of the city's two weekly newspapers, carried a capsule account of the Stanley vehicle in the making. Attributing ownership to both brothers, the *Journal* article reiterates F.E.'s optimistic expectations for his first model:

> The motor carriage which is being constructed by Mr. P.A. Murray for Messrs. F.E. and F. O. Stanley is nearly completed, and will be seen on the street in a short time...It will seat two persons, and is the lightest and most graceful motor carriage ever constructed. Its total weight ...will not exceed 400 pounds...The body of the carriage is varnished a dark green, with a handsome red blind, which relieves the monotony.

To Stanley's disappointment, he was not able to build as lightweight a car as he had described to Augusta and the local press. The two-cylinder Mason-built engine alone weighed 400 pounds, and the boiler he purchased from the Roberts Iron Works weighed over 200 pounds. According to an interview with William Mason at his works in Milton,

View of P. A. Murray's carriage shop in Newton Corner looking west on Washington Street.

Newton History Museum, Newton, Mass.

the Stanleys did not use the Regulator engine supposedly designed by Whitney but had Mason build four experimental engines for them between May and July of 1897 (*American Machinist* 1898, 896-26).

Dissatisfaction with both engine and boiler would lead to some Stanley innovations on the next model, but for the moment, the brothers had to test this first automotive effort. In late summer or early fall, the twins took the new "buggy" for its maiden run with F.O. chronicling the event: "Our first car having the Mason engine, was completed in September, 1897. I shall never forget our first ride" (Derr 1932, 46). He told the story in hyperbolic style, spinning a tale enhanced by the storyteller's art to turn in the end on a bit of Yankee shrewdness:

> We went out our alley-way onto Maple Street, and turned towards Galen Street. A horse hitched to a produce wagon was standing headed towards Galen Street. He heard the car coming, turned his head around, took a look, gave a snort and jumped so quickly that he broke the whipple tree, but did not move the wagon. He ran out to Galen Street, turned around, took one more look, then raced up Galen Street, through Newton Square and did not stop running till he reached Newtonville Square. (That afternoon the horse's owner appeared at the dry plate factory office to demand $25.)...(W)e said if he would take his harness and wagon up to Murray's and have them repaired we would pay the bill. The bill amounted to two dollars.

The *Newton Journal* for September 27 reported their test drive in glowing but decidedly less colorful terms. The paper reported that the "Stanleys' carriage was seen on the street this week and presents a splendid appearance." Although the article makes no mention of an incident with a runaway horse, it does note that the Stanley automobile presented "a novelty in Newton." The *Journal's* rival weekly, the *Newton Graphic*, however, did report an incident with a runaway horse in its issue for September 10. Though no motor car figures into the account, the story has a familiar cast. According to the *Graphic*, a horse galloped up Watertown Street, turned off into Faxon Street and crashed its express wagon into a signpost. Watertown and Galen Streets intersect at the

"Twin-powered" single engine: the Stanleys pose in their first steam car.
Stanley Museum Archives.

Watertown Bridge not far from Maple Street. Could the *Graphic* piece be the unadorned version of or perhaps the inspiration for Freel's horse tale?

No sooner had F.E. proved that his automobile could leave the hitching rail, than he began tinkering with the original design. He was determined to make a lighter car by drastically reducing the weight of the engine and the boiler. At this point F. O.'s collaboration is strongly evident. Likely on his brother's recommendation, Frank contracted with J.W. Penney from Mechanic Falls, Maine, to produce an engine that weighed only a fraction of the original Mason. Penney had built machinery for Freel's drawing tool business, and doubtless it was F.O.'s idea to contact his old acquaintance about the engine. Penney came through delivering an engine that weighed less than thirty-five pounds. The Stanleys decided to work up their own boiler design.

To lighten the boiler, the Stanleys relied on their own ingenuity and came up with a ninety-pound masterpiece. They made the shell and tubes of copper, then distributed approximately three half-inch tubes per square inch, each tube serving as a stay bolt to resist boiler pressure. But the ingenious feature was the use of high tensile steel wire wrapped in three layers around the shell, thus making a light boiler very strong (Derr 1932, 45-46). With these modifications to engine and boiler, the second automobile came very close to his original specifications—the new Stanley weighed just 400 pounds.

It is not known when this second model was ready to test. F.O. gives no clue in his 1930 account. Further confusing the issue is an interview (probably with F. E) that appeared in the December 1, 1898, issue of the periodical *American Machinist*. The reporter quotes Frank as saying that the Stanley carriage was begun in July of 1897 and finished in October of 1898. Presumably, the sequence of events subsumed in this statement is that Stanley began work on a steam car in 1897, tested it, modified its engine and boiler design, and presented the lighter weight results to the world as Stanleys' first steamer in the fall of 1898.

Whatever the timeline for building the more successful model, work had to have been substantially completed by July 1898, because by then a third automobile—a duplicate of the lighter weight steamer—was in assembly. Freel revealed this information in a letter dated July 9, 1898, to his wife Flora, vacationing in Lake Placid, New York. He regales her with the events of one of Frank's longer road tests from his home in Newton to Poland Springs, Maine. Determined to break the bicycle speed record held for this distance, F.E. embarked on a hair-raising run punctuated by several blown tires and climaxed by the breaking of the steering bar. The latter event caused the car to carom off a ledge and launched Frank, uninjured, into the bushes. Freel concludes his account with the deadpan assertion: "Had he met with no accident he could easily have reached...Poland Spring by 5:00 PM." F.O. then notes gleefully: "By the time [I] get home my carriage will be [ready] and we will have some fun"(Stanley Museum Archives).

This letter verifies that Frank built a car for himself first but controverts Raymond's contention (in the previously noted letter to F.H. Elliot) that F.O. Stanley "evidenced no interest in this first car or the next few that F.E. Stanley built…" Nor does it give credence to Raymond's suggestion that his uncle joined in the steam car venture only after he realized "what a tremendous amount of public interest the horseless carriage aroused…"—in other words when he saw the commercial possibilities of the motorcar.

Although Raymond Stanley's insistence that his father deserves sole credit for the development of the Stanley steam car is understandable and in some ways convincing, filial loyalty led him to ignore the key factor in the Stanley brothers' mechanical creativity: whenever circumstances permitted, the twins preferred to "cipher" together. They did their best work—from the construction of a mill on the family stream to the design of dry plate coating machinery—when they could bring their two heads to bear on a project.

Mechanics Hall as it looked in 1906 on Huntington Avenue, Boston. It was demolished in 1959 to make way for the Prudential Center.
1906 Boston Auto Show Program, courtesy Fred Roe.

CHAPTER 7

Mr. Stanley Builds a Winner

IN ITS OCTOBER 1898 ISSUE, *The Horseless Age* proclaimed that "a motor vehicle industry already exists in the United States." In November the journal asserted: "All signs point to the city of Boston as the first to take up the motor vehicle in the United States." Proof cited for these pronouncements was certainly not the number of autos on city streets—almost none—but rather the enthusiastic response to the first invitation issued to automobile makers to display their wares at the Mechanics Hall Exhibition in Boston. A handful of gasoline, electric and steam car companies from home and abroad agreed to participate in the 20th Triennial Exhibition of the Massachusetts Charitable Mechanics Association. Among the entrants were a large electric phaeton and small tricycle built by Riker of New York; an Eaton electric model and an internal-combustion Haynes-Apperson, both built in Boston; two Whitney steam models; three gas-powered tricycles and a four-wheeled DeDion from France.

The exhibition ran from October 10 to December 3, 1898, pulling in full crowds all the while. At the beginning of the show, the Whitney Motor Wagon Company of East Boston was the top draw, with George Whitney's steam buggy the center of attention. Whitney's car attracted such enthusiastic crowds of gawkers, many performing tactile examinations, that a distressed Whitney complained to the show organizers, who were nevertheless unwilling to restrict access to paying customers (Bacon 1984, 73).

According to Bacon's account, Whitney was most upset by the behavior of the Stanley brothers who he claimed were taking measurements and photographs of his car. But either Whitney or Bacon appears to have confused two separate exhibitions. One year earlier in February 1897, Whitney's steam carriage had been on display at the Mechanics Hall as the only motor car in a bicycle exhibit. On that occasion several eyewitnesses noted the thoroughness of the Stanley Brothers in examining the Whitney car. Henry Howard, who purchased a motor carriage from Whitney, wrote to his son-in-law in a letter dated March 17, 1948, that F.E. Stanley got "his first ideas about an automobile" at the bicycle exhibit of 1897 (Stanley Museum Archives). Nathaniel Cooledge, foreman of Whitney's East Boston shop, described the brothers' eager interest at that 1897 event: "The Stanley Bros, big, tall, bearded fellows like the Smith Cough Drop Bros... were all over that car and under it, too, measuring" (Smithsonian Archives). It is possible that Whitney was so irked by what he deemed the Stanleys' pirating of his invention that he simply repeated the tales of their earlier antics when Bacon interviewed him about the Mechanics Fair of 1898. Bacon himself may have taken some liberties with the order of events. At any rate, it seems unlikely that in October 1898 the Stanleys would have been so deeply engrossed in Whitney's car. By then, Frank Stanley was ready to try out his own steam carriage on the public.

The editors of *The Horseless Age* knew about F.E. Stanley's automobile experimentation and considered his work significant enough to be included in an article cataloging "Motor Vehicles in the United States To-day." Written as a preview of the Mechanics Hall exhibition, the report appeared in the October 1898 issue. Even though Frank was not an official entrant, his new two-passenger car was reviewed along with Whitney's and the rest. The article characterized F.E. as "a sort of all-around mechanical genius...quick to perceive the merits of the motor vehicle when the subject received its first advertisement here..." (*Horseless Age* 3 (7): 44).

The author of the article noted that the Stanley vehicle was "remarkable for its lightness": "It is of bicycle construction throughout, having ball bearings, [and] a tubular truss frame of great strength." The Stanley carriage sported twenty-eight inch wire wheels with wooden rims and two-inch pneumatic tires. The four-horse-power steam engine weighed just 32 1/2 pounds. The author was particularly impressed with the Stanley boiler, which he affirmed as one of "original construction." The author had high praise for the Stanleys' use of wire wrapping which made it possible for the boiler to safely withstand pressures as high as 200-300 pounds. Pressure in a typical automobile boiler rarely exceeded 125 pounds. The author also remarked upon the "novel design" of the gasoline burner, which could vaporize ten gallons of water with one gallon of gas.

Even though Stanley had not officially entered his automobile in the Mechanics Fair, he would get to show off his spruce little vehicle in the outdoor phase of the exhibition as an unlisted contestant. According to F.O. Stanley, this came about after the sporting editor of the *Boston Herald* spoke to Isaac Davis, organizer of the exposition. Davis, no doubt pleased to expand the ranks of the participants, issued an invitation to Frank. The car Stanley would drive in the trials was the same one he had battered on the wild ride to Poland Springs. Currently, it was the only model the Stanley's had. A month before the Mechanics Hall show opened, Freel sold its twin to one E.T. Ebens Methot, a dentist from Dorchester, for $600 (Derr 1932, 47).

The open-air portion of the Mechanics Exhibition—the field competitions—took place on November 9, 1898, preceded by an automobile parade from the hall on Huntington Avenue to the Charles River Park Velodrome in Cambridge. The Riker Phaeton, which competed in all but the speed trials, led the parade, and though his name did not appear on the program, F.E. joined the procession. When he entered the track unheralded in his "light and trim steam carriage," he got a "spirited" (if mystified) response from the crowd of 1500 onlookers (*The Horseless Age* 1898, 14).

Stanley did not compete in what was anticipated to be the big event of the day—a two-mile pursuit race. The gas-powered DeDion tricycle from France was considered the favorite, but race officials were hoping for a tight competition. The DeDion did finish first with a time of 5 minutes, 1 2/5 seconds; Whitney's steam carriage took second and no one else was close. *The Horseless Age* lamented that the official race had been "a tame affair," with only Whitney and the DeDion in contention. At that they noted the performance of the winner "was not at all brilliant." Indeed, the entire event was most memorable for the "vile odors emitted in abundance" by the DeDion tricycles (*American Machinist* 1898, 21-971).

The razzle-dazzle came at the end of the day and belonged to Frank Stanley who participated unofficially in the hill-climbing trials. From a level running ten-foot start, each car was to drive up an 80-foot ramp canted to a maximum 36 1/2 per cent grade. The DeDion tricycle never got higher than forty feet in either of its attempts. Whitney sealed the event among the official entrants in his first effort with a climb of 78 feet in 6 4/5 seconds. Then, as Nathaniel Cooledge saw it, "the gates opened and in came Stanley with a snappy little car that flew around the track and up that ramp" (Bacon 1984, 77). To the roaring approval of the crowd, Stanley first circled the veladrome three times. With only a thin grayish plume of odorless exhaust trailing behind the steamer, he drove a mile in the fastest time of the day—two minutes, eleven seconds. Then, in one try from a dead stop, Stanley made it all the way up the ramp, bumping the crossbar at the top in the fastest time of 4 3/5 seconds. Though Whitney came within a hair of eighty feet on his third attempt in the official winning time of 5 2/5 seconds, Stanley's performance carried the crowd.

Perhaps this was the most galling event of the entire exposition for Whitney. All the popular interest that had centered upon his steam car in the indoor phase of the exhibition simply dissipated. Now it was Stanley who drew the excited crowd lingering to ask him questions long after the time trials were over.

Buzz from the exhibition translated for the Stanleys into requests for their autos. Within two months of the Mechanics Fair Exhibition, the Mason Regulator Company reported that it was filling an order of

Charles River Park, Greater Boston.

This Bert Poole illustration shows the bicycle track at Charles River Park, scene of F.E. Stanley's first steam triumph.

Stanley Museum Archives.

one hundred engines for the brothers (*The Horseless Age* 1898, 22). By February 1899, the Stanleys had taken on eighty-five more orders (*The Horseless Age* 1899, 18) and had been offered a multi-year contract to supply their cars to a consortium of European investors. Ready or not, the twins had fallen into the car business. Since they were still fully engaged in the manufacture of photographic dry plate, they produced their steam autos under the Stanley Dry Plate corporate name and at first even assembled the cars in their dry plate factory.

According to Raymond Stanley, the brothers had to make all parts themselves or piece the work out under their supervision to neighboring machine shops. They ordered their auto carriages from Currier and Cameron (later called Shields Carriage Company) which sent bodies equipped with leather dash and whip socket—the latter primarily for show. Once again, the Stanleys had their engines built locally by the William B. Mason Regulator Company of Milton.

On April 12, 1899, a brief bulletin in *The Horseless Age* announced that the Stanleys had purchased the Old Hickory Wheel Factory in

Watertown. This had been Sterling Elliott's bicycle shop, located just across an alley from the Stanleys' dry plate factory. Here Elliott had assembled his "Hickory" bicycle, but his business failed in 1897, and he now devoted his full time to publishing and editing a bicycling magazine for the League of American Wheelmen. F.O. Stanley bought the contents of Elliott's machine shop at auction, and in the spring of 1899, the twins purchased the vacant plant for "a fraction of its cost." (Derr 1932, 48).

With a factory building now devoted to the manufacture of automobiles, the Stanleys confidently anticipated large-volume production. Echoing the optimism of the brothers themselves, *The Horseless Age* proclaimed that the Stanleys "are fitting up to 1000 carriages per year." Even with the new factory, the brothers' production plans, which far exceeded anything attempted by an American automobile maker up to that time, proved overly ambitious. The brothers soon had difficulty keeping up with their orders. One anecdote relayed from this period tells of an exasperated customer writing the Stanleys that he was afraid he might die before his car arrived. The twins reputedly issued him the Yankee version of the customer is always right. They told him not to worry; they would "upholster {his} car in asbestos" (Villalon 1981, 17).

The difficulties of meeting their production schedule did not stop the brothers from automotive innovation. Late in April, *The Horseless Age* reported the Stanleys had assembled a new model—longer in the body with redesigned and improved running gear and suspension. F.O. described this latest Stanley as "much better than any we had made" (Derr 1932, 49). He himself was now driving the prototypical version as a replacement for the car that he had sold in the fall. This improved model was the same one the brothers were now producing for their customers. By late spring, the first of these new Stanleys were ready for delivery. That announcement came in the May 24, 1899, edition of *The Horseless Age*. The news reported on June 7 was clearly more startling: the Stanleys "have sold their auto carriage business to John Brisben Walker." This surprising announcement, however, was only half the story. Walker would have to take on a partner to clinch the deal.

CHAPTER 8

A Stanley by Any Name

JOHN B. WALKER was the owner and publisher of *Cosmopolitan Magazine* and an early aficionado of the horseless carriage. In 1896, inspired by the Chicago "motocycle" race a year earlier, he sponsored a similar event in New York City. Entrants were to drive from the Manhattan Post Office to Cosmopolitan Headquarters in Irvington-on-the-Hudson and back into the city—a distance of fifty-two miles. New York City officials reluctantly permitted the race to proceed with the stipulation that contestants not speed until they were beyond the city limits. The drivers, however, ignored that restriction and a mad dash ensued through Manhattan (Villalon 1981,12-14). The race came off like a Barnum and Bailey show with competitors rattling and careering over treacherous cobblestone pavement in a desperate effort to avoid collisions with horse carriages, cable cars, and Civil War veterans dispersing after the Decoration Day parade (Bacon 1984, 65-70). Only the Duryea brothers (victors in the Chicago race) managed to return to the city on the same night.

The clownish outcome of the New York event did nothing to dampen Walker's enthusiasm for the automobile. In 1898 he attended the more somber Mechanics Exhibition in Boston and marked the impressive Stanley performance in the Charles River Park competition. According to F.O., Walker first approached the brothers in February 1899 with an offer to buy half interest in their automobile business. Freel recorded in bemused style how the brothers allowed they did not have "what might be called an automobile business" and that they "certainly did not want a partner" (Derr 1932, 48). Walker persisted, however, and returned to the Watertown factory in early April with the intent to buy the brothers out. The Stanleys set a price they thought so high—$250,000 in cash— Walker would go away. To their astonishment, he accepted the asking price without a quibble and left a $10,000 check on a ten-day option to raise the quarter of a million dollars.

In order to have something tangible to show potential investors, Walker telegraphed F.O. Stanley to ask if he would ship the steam car from Providence, Rhode Island, to New York City by ferry. The chance for a bit of automobile theatrics apparently tickled Stanley who gladly made the journey with the little steamer. By prearrangement with the captain of the boat, Freel got to drive the car off the ferry. Shortly before the boat docked, he fired up the boiler on the Stanley so that he might disembark before anyone else. F.O. rolled down the gangplank to the rousing applause of fellow passengers (Derr 1932, 51).

His friend Sterling Elliott joined him on the docks after arriving in New York on a different ferry. The two then set out upon what Freel called "the most perilous journey I had ever undertaken in my life." They intended to drive the car from the docks to Walker's office on the Hudson. As soon as they turned into the street in front of the wharf they were engulfed by traffic. The roadway was teeming with vehicles of every description—all of them horse-drawn. Stanley managed to steer safely among the frightened animals and cursing drivers only to be hit broadside by a young woman on a bicycle. As F.O. told it, the impact catapulted the girl unharmed into Sterling Elliott's lap, but an angry crowd soon gathered. Luckily, the police officer on the scene was kindly disposed to the horseless carriage and chastised the young lady for her careless riding.

The expedition eventually made it to *Cosmopolitan* headquarters in Irvington where Walker had set up several days' worth of test drives for wealthy young men with cash at their disposal. Stanley first entertained the two sons of Jay Gould who professed to like the automotive experience but did not invest.

Despite Freel's misgivings, Walker himself, who curiously enough had never driven an automobile before, decided to take William Rockefeller (brother of John D.) on a test drive. After Walker spent a few minutes getting accustomed to the steering bar, the two set out for Tarrytown. Stanley then learned from Walker's son that overnight and unbeknownst to his father, the car had developed a small leak between the water tank and the pump. Since young Walker had closed the valve when he detected the leak, this meant that his father was driving with no water going into the boiler.

To prevent the catastrophe of a scorched boiler, Stanley telephoned ahead of the pair to a drug store along the way. F.O. no doubt had scouted out this apothecary when mapping routes for the demonstration runs. In the days before gas stations, automobilists could purchase gasoline from drug stores, and the steam-powered Stanley needed gas for its burner. Stanley's desperate call persuaded the druggist to stand outside his establishment and flag down Walker as he came by. With emergency instructions from Freel, Walker managed to fix the Stanley, and he and Rockefeller traversed the seven miles to Tarrytown unscathed. Despite his heroic efforts, (or perhaps because of them) Walker could not induce Rockefeller to invest in the steamer business. Deeply shaken by the day's events, Walker decided to leave the driving to Stanley.

By F.O.'s telling, it was on the very last day of the ten-day option that Walker secured a financial partner. He asked Amzi Lorenzo Barber, the "asphalt king" of New York, to take a ride in the Stanley. Barber, like Walker, was one of the first American businessmen to comprehend the importance of the horseless carriage. He had a monopoly on the asphalt beds in Trinidad and quickly perceived his paving fortunes linked to the nascent automobile industry (Buffalo and Erie County Historical Society). In 1900, that industry was so new that it was not yet clear which engine technology—steam, electric, or gasoline combustion— would dominate future development. But Barber was trying hard to figure it out because he intended to go into the car business. Ever since visiting Whitney's shop in East Boston in the mid-1890s, he had been leaning towards steam-powered cars.

The test drive in the Stanley apparently made up his mind. After a long session with F.O. at the steering bar, Barber agreed to stake Walker to the $250,000 for a half interest in the business. Just how high a price Walker really paid to procure Barber's backing became evident in a matter of weeks. As for the Stanleys, they sold Walker and Barber both the Old Hickory Wheel factory and their patent applications on a tubular steam boiler and a steam generator. The brothers then signed a contract to serve as general managers for the new company, with the stipulation that they themselves refrain from manufacturing steam cars until May 1, 1900.

The June 21, 1899, edition of *The Horseless Age* announced the Walker-Barber alliance, incorporated under the laws of West Virginia as the Automobile Company of America with capital of $2,500,000 (John D. Rockefeller himself was now a heavy investor). Amzi Barber, John Walker, and Samuel T. Davis, Barber's son-in-law, headed the new corporation. The article also mentioned that another automobile company by the same name had been incorporated seven months earlier in New Jersey. To "avoid confusion," the reporter speculated, "the name will be probably changed."

Sometime between the corporate announcement and the second week in July 1899, Barber and Walker did indeed change the name of their enterprise to Locomobile Company of America. It was under the brand name of Locomobile that most Americans became familiar with the first Stanley-made cars. The "Stanley Horseless Carriage" offered for sale by Locomobile was a heavier, more roadworthy version of the model that Frank drove at Charles River Park. It took only about ten minutes to get the pressure up in the boiler; thereafter, the little car, now equipped with a reverse gear for the operator, could reach speeds of 40 miles per hour and travel (by some reports) a hundred miles before refueling. In fact actual mileage depended on the experience of the driver. As one exasperated doctor wrote the editors of *The Horseless Age* in the September 13, 1899, edition: "The gasoline tank holds a little less than 3 gallons, and with that quantity of oil they (the Stanley Brothers) can run the carriage 75 miles, but I have not heard of anyone else who can do it yet." The doctor's own mileage after one week's driving was 16 miles on 1 1/2 gallons of gasoline and three quarters of a tank of water.

Still, compared to the contemporary gasoline cars, which smelled bad and were exceedingly noisy, the Locomobile ran silently, smoothly, and dependably. As the first Locomobile advertisements proudly proclaimed, once the driver mastered the simple operation of the vehicle, the motor carriage demanded only about one-twentieth of the time needed to care for a horse. Walker and Barber hoped to produce ten of these cars a day in their Watertown factory and offered them to the public for the appealing price of $600.

Despite its popular product, the new company did not make it to its first month's anniversary. The trouble, however, lay not in the cars but in the men. The July 19, 1899, issue of *The Horseless Age* announced a parting of the ways for Barber and Walker. The grounds for the split remain unknown, but it is evident from the terms of dissolution that it was a nasty divorce: Barber got to keep the Locomobile name and the Watertown factory. Walker, though compensated monetarily for his share of the property, had to start all over with a new company name and a new plant. He called his firm the Mobile Company of America and planned to manufacture his Stanley steam vehicles in a factory on the Hudson. Walker was to produce Stanley cars, but his legal relationship to the Stanley patents was apparently terminated. At least that was the opinion of Charles Foster, the twins' patent attorney, as related by Frank in a letter dated February 22, 1900, to his brother Freel:

> Foster says that Walker has no claim on us. That according to the conditions of the contract we had with him unless he paid the $240,000 on or before the 4 (sic) day of August he forfeited his options and when we agreed to accept Barber's payments in

place of Walker's we surrendered our claim on Walker and his claims on us (Stanley Museum Archives).

Although his exact status regarding the Stanley patents cannot be stated for certain, Walker was clearly on the short end of the deal. The loss of the popular "Locomobile" name alone was a significant setback; indeed, Walker struggled for many months to get his infant Mobile operation going. It was March 7, 1900, before *The Horseless Age* announced that Mobile Company of America had completed "its first carriage at the Kingsland Point factory." The facility at Kingsland Point was a large one, capable of turning out 600 cars per month, but Walker's operation never lived up to its production potential. By October 4, 1900, the automobile journal reported that the Mobile Company had discharged all but 100 of its 560 employees.

For his part, Barber hardly missed a step in production. He added his son, LeDroict Landon Barber, to the company management and immediately announced the purchase of an auxiliary factory in Westboro, Massachusetts, then the leasing of another plant in East Worcester. Business continued to be so good that in July of 1900, Barber made plans to build a factory on a large tract of land he had purchased at Bridgeport, Connecticut. Locomobile soon became the largest manufacturer of automobiles in America (Villalon 1986, 17).

The impact of events on the Stanleys, however, was surprisingly minimal. Their automobile would now be manufactured under two different brand names and the brothers, like children of divorce, expected to spend equal time as general managers of both companies, at least for the duration of their one year contracts (*The Horseless Age* 1899, 12). Divided loyalties did not keep the twins from demonstrating the superior nature of their automobile—regardless of whose name was on the label—to anyone who cared to watch.

CHAPTER 9

Out and Back in Again

AS GENERAL MANAGERS for Locomobile (and later Mobile) the Stanley brothers' primary function was to find markets for the steam car. Before the turn of the century, that task was harder than it seems, because the public, while enjoying the novelty of horseless locomotion, did not yet know how seriously they intended to take these motorized contraptions—steam, gas, or electric. While automobile enthusiasts in Paris, New York, Boston, and Chicago were rapidly forming into clubs, many other people found the expense of purchasing a vehicle and the cost of its upkeep too high. Some considered the automobile a dangerous nuisance.

In the metropolises of Western Europe and the United States, where the automobile first appeared in any numbers, civic ordinances pertaining to the operation of horseless carriages reflected a similar ambivalence. Paris, with a considerable amount of motorized traffic, already required driver's licenses for anyone operating an automobile. New York City, while embracing the electric taxicab, passed tough regulations concerning steam vehicles. Under the assumption that operating a steam car was analogous to driving a locomotive, New York made automobile operators pass an oral examination on the construction and operation of boilers (*The Horseless Age* 1899, 7-8). William K. Vanderbilt, Jr., an early automobile aficionado, was arrested on Fifth Avenue in 1899 for operating his Locomobile without an engineer's license. No cars of any design were allowed in Central Park. Yet the New York Police Department offered both driving and mechanical repair lessons on all types of automobiles.

From London came the most promising sign that the motor age had arrived to stay: all the horses in Queen Victoria's royal stables now regularly drilled in the presence of automobiles (*Scientific American* 1899, 282). This piece of news was the more significant considering that just three years earlier, Britain still had "red flag" laws that severely inhibited the operation of steam vehicles. As far back as 1831, Parliament required a flagman to precede any steam-powered vehicle on the road with a red banner by day and a red lantern by night. Repeal of these laws in 1896 was hailed literally as a green light for the steam car industry in Great Britain.

Before the turn of the century, even the nomenclature for the infant industry remained open to debate. The French coined the word "automobile" which had begun to gain some popular currency in the American press. Typical of the more fanciful entries, The *Scientific American* concocted the term "autopher," a mixture of the Greek "auto" and, presumably for travelers' luck, "St. Christopher." For its part, *The Horseless Age* took umbrage at what it called "all the fantastic philology" and opted for a practical approach. "The terminology best suited to our needs here in the United States is the motor terminology…regardless of the frantic efforts of word coiners to foist their barbarisms on the language"(1899, 6). In a few years, when their stint with Barber and Walker was at an end, the Stanley twins would choose "motor carriage" as part of the official name of their new car company.

In the meantime, they set about their sales responsibilities with characteristic dedication. The brothers divided their duties territorially between Europe and the United States. Frank spent most of his selling time abroad, trying to find a niche for the steamer in a European auto market that seemed already to favor the combustion engine. For his part, Freel drew domestic duty when his health permitted. Much as he once had taken to the road to demonstrate the superiority of Stanley dry plate,

F.O. now traveled the United States drumming for Locomobile. With his fine sense of the dramatic, he often found the biggest stage possible for his demonstration runs. On a jaunt to Washington, D.C. in August 1899, he persuaded an apprehensive and reluctant President McKinley to take a test drive, thus making him the first US president to ride in an automobile. Several presidencies later, William Howard Taft made a steam vehicle (though not a Stanley) the first official White House automobile.

Freel's travels, not surprisingly, inspired him to tell many a good story that in the repetition served to enhance the lore of the steam car. He particularly favored one tale about a train trip into the South (perhaps the same junket that took him to the White House) with a Locomobile disassembled and packed in two large crates. When he arrived at his destination, he had the crates hauled to a field where he began putting the pieces of the car together with the help of a mechanic and several local volunteers. This extraordinary sight drew quite a crowd of curious farmers and field hands. When Freel had finished the task of assembly and gotten the steam up in the boiler, a series of sharp reports emanated from beneath the Locomobile. The noise was sufficient to scatter all but one of the onlookers. The sole remaining spectator, a black farmer, was smiling and pointing underneath the car. A baffled F.O. bent over to see the joke—it seems the man had tossed a couple of firecrackers under the carriage. Ever one to appreciate a good prank, Freel shook hands with the fellow and offered him a test ride. As they turned into a toll road, F.O. asked the startled toll keeper, "Have you seen our horse?" This quip apparently hit the mark so successfully that Freel often repeated it in his travels (Derr 1932, 59).

The most widely hailed demonstration ride that Freel took for Locomobile, however, made the newspapers and became famous without the agency of the Stanley narrative flair. On August 31, 1899, with wife Flora gamely by his side, F.O. drove a Locomobile up the carriage road that ran to the top of Mount Washington, the 6,288-foot queen peak of New Hampshire's White Mountains.

Freel made the most of the event by taking a leisurely four-day drive from Newton to the base of the mountain. On August 26, he and Flora left their home on Hunnewell Avenue in a Locomobile model mechanically superior to the little Stanley that had wowed the crowds at Charles River Park. But the Locomobile still had no top (fortunately the weather en route was fine), no windshield and just room enough for two passengers. Their luggage had to be sent ahead by train.

During the first leg of the journey, the Stanleys drove fifty-six miles to Newburyport, Massachusetts. Traveling this distance required one stop to oil the exposed cranks and connecting rods underneath the seat of the carriage and another to refill the twelve-gallon water tank from a farmer's trough. On the remaining days of the trip, they drove only a few hours each day at an average speed, calculated by F.O., of 14 1/10 miles per hour (Flora Stanley [1899] 1999, 14). The slow pace was dictated in part by the poor condition of the roads, which were mostly rough and sandy. On some stretches of road, the sand was so deep that F.O. later explained that "it took the full power of the machine to drive it on a level" (*Among the Clouds*, 1 September 1899). But the leisurely rate of travel was also determined by the more pleasurable business of bringing a motorized novelty to folks along the route.

In the summer of 1899, the sight of a horseless carriage was such a rarity in rural New England that the passing of the Locomobile astonished both equine and human populations alike. Young boys were often the first to espy the auto. In the time it took these little sentinels to alert the other denizens of the farm, Freel and Flora had passed quietly by, thus prompting Flora to muse on the fate of these fellows: "I have often wondered how many small boys underwent chastisement for lying, or treatment for disordered nerves." Whenever the Stanleys stopped at an inn to eat or a hotel to sleep, their arrival drew an excited, and as Flora noted, democratic crowd, with "landlords, waiters, and guests hobnobb(ing) equally..." (Flora Stanley 1999, 12, 14).

When they finally reached the mountains in New Hampshire, the Stanleys rested a whole day at the Kearsarge in North Conway before driving on to Darby Field Cottage at the base of Mount Washington. They spent the night at this cottage named, appropriately enough for the occasion, after the first white man to climb the peak in 1642. The next morning the couple got an early start on the five-mile drive to the actual jumping off point for their adventure, the site of Glen House. Located in Pinkham Notch on the eastern slope of Mt. Washington, this grand hotel had burned down for the second time in 1893 and never been rebuilt.

The Stanleys had decided to make their ascent in the morning to avoid afternoon teamster traffic on the mountain. The last thing the fledgling Locomobile company could afford to do was send a terrified team of horses over the side of a mountain road.

To prepare for the climb, F.O. removed the chain drive from the car and examined it link by link. He cleaned and oiled each link separately before reinstalling the chain. "Then feeling assured that we would not be called upon to use the brakes which were not too dependable, we proceeded on our way" (Derr 1932, 55-56). Perhaps his friend Sterling Elliott convinced him that numerous bicycles and tricycles, with equally primitive brakes, had successfully negotiated Mt. Washington. Besides, on the way up, brakes did not concern Stanley as much as making good time. To that end, he decided to keep the steamer light by filling the water tank only half full. He intended to stop for a refill half way up the mountain.

Thus, at nine a.m., semi-watered, well oiled, and fingers crossed, the couple turned onto the Mount Washington road to follow its torturous route to the summit. To call a twelve-foot wide stretch of rock-studded, rutted sand, broken some 350 times by water bars and rugged outcroppings, a road may have been a bit of an exaggeration. Completed in 1861, the roadway was eight miles long with an average grade of twelve percent. Above the tree line, where the mountainsides pitch away steeply with drop-offs of several hundred to several thousand feet, the road was often shrouded in mist. Under good conditions, a horse-drawn stagecoach could make the ascent from Glen House to the

Summit Lodge in four hours, but on this particular summer morning, favored with clear, calm weather, Freel and Flora would cover the route in two hours and ten minutes.

The first upward section of the climb proved so steep that the Stanleys had to stop three times to jack up the rear wheels of the car and run the engine for a few minutes in order to get sufficient water to the boiler. The stops delayed them by about twenty minutes. No doubt as a result of Freel's mountain experience, the 1900 Locomobile model was equipped with a hand-operated pump to supply water to the boiler. The following summer, a young Newtonite named Joseph W. Crowell took on Mt. Washington again and, driving the latest model Locomobile with the hand pump, managed to shave six minutes off the Stanleys' time (*Among the Clouds*, August 7, 1900).

As Flora and Freel made their way up the mountainside, Flora rendered a good description of their mechanical ascent: "Then we went on and up, still up, the continuous climbing varied only by a steepness so excessive that as we looked ahead to it, we felt sickening anxiety lest each brilliant dash should be our last." Though "panting and quivering," the Locomobile engine did not fail them (Flora Stanley 1999, 14).

When they reached the Half Way House at 3,840 feet above sea level, F.O. found to his surprise that he had used very little water. Assured by the housekeeper that there was a spring at the five-mile marker, he decided to press on before refilling. Then, as Flora told it, they became so engrossed in viewing the scenery and "watching each pulsation of our machine," that they simply overlooked the mileposts. They reached the seven-mile marker before realizing their mistake. While jacking up the car for the last time, Freel announced the boiler was nearly empty. So Flora agreed to scout for water on foot. To her relief she soon encountered a teamster who apparently had been sent down to check on their progress and now willingly returned to the summit for water. After a break to refill the tank, the Locomobile made it easily to the top.

Telephone wires were down that day on the mountain, so news of the Stanleys' impending arrival traveled by word of mouth to the Summit

Without a windshield, without a top, and certainly without seatbelts, Flora and Freel made it unscathed to the top: "…(F)or a gentleman and his wife to ride up Mt. Washington by such a vehicle will cost less than 25 cents." Photograph by Frank Burt.

Stanley Museum Archives.

House, the lodge that once stood lashed by wire cables to the top of Mt. Washington. Some fifty guests who had traveled by cog railroad for the occasion now crowded onto the end of the rail platform or stood atop the hotel woodshed to catch a view of the finish. For his part, Frank Burt, the local editor covering the event, did not have far to go to get his story. The offices of the vacation-time newspaper, *Among the Clouds*, were located a few steps away in a second summit lodge called Tiptop House. Since Burt was a friend of the Stanleys and a fellow Newtonite, this was a proud as well as newsworthy day for the young man.

When Freel and Flora reached the summit, Frank Burt recorded the moment for the September 1 edition: "'Here they are,' was passed from lip to lip, {as} the little vehicle with its two passengers was seen rounding the curve." The Stanleys obliged the cheering onlookers, as well as Burt's camera, by posing for photographs in their two-seater. Then as they alit, F.O. remembered his mission: "This is the regular Stanley Locomobile," he said. "The amount of gasoline required to ascend the mountain was less than two gallons. Thus it will be seen that for a gentleman and his wife to ride up Mount Washington by such a vehicle will cost less than 25 cents."

The next morning, the descent proved much easier than the climb. Any doubts Stanley may have had about the braking ability of the Locomobile quickly dissolved. In Flora's words, "We reversed the engine, played the brake like an organ pedal, just held on and let the thing spin" (Flora Stanley 1999, 16). They safely reached the bottom in one hour, including a stop off to pay the $1.92 toll at the Half Way House.

Though New Hampshire locals paid scant attention to the doings of city folk on the mountain that day, the Stanley climb made the metropolitan papers both at home and abroad. News of the ascent was cabled across the Atlantic and printed that same evening in the Paris edition of the *New York Herald.* As a publicity tour, the Stanley climb found its intended mark. "This car climbed Mount Washington" proved a successful advertising slogan for Locomobile long before it became a

The Tally Ho Coach: eight-horse-power ride to the summit. The Stanleys did it faster and with fewer "horses." Photograph by Keith's Theatre.
Splash Pan, *Stanley Museum Archives.*

bumper sticker for White Mountain tourists. International reporting of the triumphant climb may also have given Frank's sales efforts in Europe a final boost since word came just as he was winding up his first summer abroad.

Almost as soon as the Walker-Brisbane deal was sealed, Frank had set out for Europe. He sailed on June 28, 1899, in his capacity as general manager of one company and returned the following September as manager for two. The change in company status, however, seemed to have little bearing on his activities in Europe.

Upon his arrival in France, Frank was convinced of the great marketability of his little car. He wrote his brother in a letter dated July 7, 1899, about seeing "hundreds of motorcars" in the streets of Paris. Since most of them were gas-powered DeDion tricycles, which the Stanley car had outclassed in the Charles River Park trials, Frank was optimistic their car would sell well in France. He also noticed that there were almost no steam vehicles on the French market and felt sure the Stanley stood in good position to fill that niche. "Steam," he told Freel, "is conspicuous on account of its absence." He also noted with a certain amount of envy how good the roads in Paris were: "The worst streets I have seen are almost equal to our best." He marveled that they had been constructed practically with "no grades." After viewing the 600 entrants in the Paris auto show, Frank was convinced that the Stanley machine could overtake any of them on the good Parisian roads (Stanley Museum Archives).

As of July 17, he was still awaiting the arrival of the Stanley Locomobile for display in the Paris auto show. He wrote Freel: "I am more confident every day that a machine like ours will have a larger sale here both on account of price and quiet running" (Stanley Museum Archives). Stanley's instincts proved sound. When the Locomobile model finally docked, Frank found his steamer the object of much interest in French automotive circles.

He was wooed in high style by the head of one Parisian company eager to represent Locomobile in France. Viscount des Jotemps, who had earlier negotiated a contract with the Stanleys to produce steamers for the European market, invited F.E. to dine with him at the Automobile Club of Paris. Frank attended with Gustie, five-year-old Raymond, and daughter Blanche in tow. (Younger daughter Emma was by this time married to Prescott Warren and did not accompany her parents). Having at last won her father's blessing to pursue a singing career and with her mother's sister Nell as chaperone, Blanche had preceded her parents to Europe. She was eager to display her continental polish and a bit nervous about her father's ability to move in sophisticated circles. Since F.E. sailed after Gustie and Raymond, he presumably had packed for Europe without wifely supervision. Blanche feared that on this most elegant occasion her father would appear in what she termed his customary disregard for proper dress. As she later revealed in her memoirs, the fears of his female entourage proved groundless; F.E. stepped out into Paris society "perfectly groomed." Blanche discovered that he had come abroad with a handsome new set of luggage filled with "tasteful suits for all occasions" and shoes that "were a credit to the finest American cobblers" (B. Hallett 1954). But there were even more surprises in store as the evening unfolded.

The viscount whisked the sartorially correct Stanleys to the Automobile Club of Paris, which was situated at a corner of the Bois de Bologne and had a dining room so elegant that Blanche felt she "had awakened at the court of Marie Antoinette." There the family was treated to a lavish meal presented in multiple courses, each accompanied by a separate wine. To the utter astonishment of Blanche and her mother, Frank—the lifelong teetotaler—sampled each wine as it came. After dinner the entire party repaired to the Cafe de Paris for a taste of the city's vaunted nightlife. More wine and more revelations flowed. By his daughter's account, F.E. kept bibulous pace with his host and betrayed no ill effects from the wine.

For the remainder of the Stanleys' time in Paris, Frank adhered strictly to the local custom. Each evening, after the business of the day was over, he dined with his family and insisted on having a fine wine "with every meal." In the manner of a French papa, he even let little Raymond sample the wine. When they sailed for home, however, F.E. mostly left the wine in Paris and apparently his eye for dress as well. Some seven years after that trip, Gustie confided to her diary in dismay that Frank had dressed for a dinner party in an outdated suit that she thought had long since been put up in the attic.

By all business measures, however, F.E. returned to the United States in good fashion. He sailed for home at the end of August with four hundred orders for the Stanley No. 1 Auto. This was the basic Locomobile sturdier and with better suspension than the carriage Frank drove at the Charles River Park exhibition. By the time he returned home, Locomobile was offering its customers a buggy with a phaeton top and had plans to bring out a four-seat model in two months.

The summer of 1899 proved to be a busy one abroad for other American steam automobilists. While F.E. Stanley was seeking European sales for Locomobile, old friend George Whitney was displaying his latest automobile model in London and Paris. An article datelined July 26 in *The Horseless Age* described Whitney as "the hero of the hour" at the London Exposition and "expected him to take several prizes" there. Even the Stanleys' boss, Amzi Barber, personally joined the overseas competition. In its August 23 issue, *The Horseless Age* reported that Barber and his chief engineer Herbert Searles were in England to establish a Locomobile branch. The article could have added parenthetically that Barber was also on the lookout to buy the American manufacturing rights to any available European steam car patents.

Sometime in late summer, Barber crossed paths with Whitney. The two men were not strangers to each other since Barber had been an early visitor to Whitney's East Boston shop (Bacon 1984, 63). Indeed, the Whitney car had gone a long way towards convincing Barber that steam was going to be the dominant technology in the automotive industry. Now in Europe, Barber was so impressed by the new Whitney carriage that he made preliminary overtures to buy up Whitney's patents. At the time, however, the Whitney Motor Wagon Company was manufacturing automobiles through several licensees—the Stanley (no connection) Manufacturing Company of Lawrence, Massachusetts, the Anderson Manufacturing Company of South Boston and Brown Brothers Ltd. of the United Kingdom. Before accepting any offer from Barber, Whitney would have to settle accounts with his licensees.

The exact date of the meeting with Whitney is not known, but John B. Walker did not figure into the talks. Barber made the proposal to Whitney in the joint name of Sam Davis, his son-in-law and co-officer in the new Locomobile Company. No deal was signed in Europe, but Barber's offer was so attractive that Whitney made plans to break the contracts with his licensees when he got home. For his part, Barber insisted that his mechanic subject the Whitney carriage to a thousand-mile road test before the final signing (Bacon 1984, 87-88).

When all the impediments were finally cleared, the deal closed February 1, 1900 (Bacon 1984, Appendix D). Barber agreed to pay Whitney $290,000 in cash and stock options for the Whitney Motor Wagon Company and sole rights to its patents (Bacon 1984, 87). That same month, Whitney took over as chief designer of the experimental division at Locomobile. All of these events occurred before John Brisben Walker had produced a single car at his new factory.

Interestingly enough, Walker was not completely out of the picture. In July 1900, The Stanley Automobile Company of New York City was incorporated with capital of $5000. Barber created it solely as a repository for the Stanley patents and a holding company for the acquisition of additional automotive patents (*The Horseless Age* 1900, 24). The directors of the corporation were Amzi L. Barber, Sam T. Davis, and somewhat surprisingly, John B. Walker and his son David S. Walker. Walker's inclusion on the board of directors suggests either he still had some sort of claim on the Stanley patents or he was so desperate to stay in the steam car business that he would risk another venture with Barber.

As for the Stanleys, they now found themselves working with George Whitney for the same boss. This must have been an awkward arrangement for all concerned, considering how strained relations among the three of them had become since the Mechanics Hall Fair. Whitney came away from that event utterly convinced that the Stanleys were stealing his designs. For their part, the brothers may have had their feathers ruffled by the installation of Whitney as chief designer for Locomobile.

On the other hand, the Stanleys had never been in any sense "company men." Even when carrying out their duties as general managers for Locomobile, they did so with an air of independence. F.E.'s letters to his brother during his summer abroad always refers to "our car" but the ownership implied is between the twins, not with Amzi Barber. One senses that the brothers complied with the terms of their deal but, when their year commitment ended, gladly walked away. The arrival of Whitney at Locomobile may have made their departure that much easier.

By the terms of their contract with Barber, the Stanleys were prohibited from starting another auto business until May 1, 1900. It is probably around this time that they left Barber's operation. Several months earlier, correspondence between the brothers revealed they were preparing to leave Locomobile. In the previously cited letter of February 22 from Frank to Freel—which posited no existing contract between Walker and the Stanleys—another interesting bit of information emerges: Barber has fallen in arrears on his contract payments to the Stanleys. Frank, however, does not seem too worried about default: "I am satisfied now that Barber will settle...He is amply able and has got any amount of property in his own name." Thus hopefully convinced that Barber will pay up and that Walker has no viable claim on them, Frank concludes, "we must make our plans accordingly" (Stanley Museum Archives).

There is no evidence that the Stanleys began to manufacture automobiles as soon as they finished business with Barber. The first public notice that the brothers had plans for a new car model came a full year later, on February 13, 1901, in the pages of *The Horseless Age*: "It is said that the Stanley Brothers, Newton, Massachusetts, inventors of the locomobile, intend to put out a lighter and cheaper carriage in the Spring."

While under contract to Barber, however, they clearly had continued to improve upon their automotive technology. As F.O. stated in his 1930 retrospective: "During the interval between 1899 and 1901 we were not idle" (Derr 1932, 53). In fact they had applied for several new patents that had been assigned to Amzi Barber and Locomobile. Since these applications were not part of the original sale to Barber and Walker, Barber acknowledged the mistake and agreed to let the brothers assign the patents to whomever they wished, pending his approval.

After their contract with Barber was concluded, the Stanleys simply kept on working. From December 11, 1900, to July 23, 1901, they were granted patents on running gear, an engine, a superheater, and a flash boiler (never used on a Stanley, but leased to the White Company). On March 2, 1901, the brothers filed jointly for a motor vehicle patent—their first on any model. Three months later on June 4, 1901, their application was granted. By the time they were ready to build cars again—more than a year after their break with Barber—they had designed an automobile, a 6.5 horsepower runabout, that was in F.O.'s phrase, "far superior to any before made."

CHAPTER 10

Whitney v. Stanley

WHEN THE STANLEYS BEGAN to manufacture the first of these new models is not known for certain. What seems clear is that the brothers could not have assembled automobiles in any significant number without a factory. Amzi Barber had bought their Watertown plant, machinery, and patent applications in 1899. Although the Stanleys' contract with Barber expired in May 1900, it was a whole year before the brothers got their factory back. To the best of F.O. Stanley's recollection, Barber sold them the enterprise—minus the original Stanley patents—in May 1901 for "the modest sum of $20,000"(Derr 1932, 53).

F.O. speculated that Barber was willing to sell the Stanley business so cheaply because he had begun building internal combustion cars in 1901 and by implication had lost interest in steam vehicles (Derr 1932, 53). But this was not quite the case. Even though Locomobile began to manufacture gasoline models in 1902, it continued to produce steamers through 1904. Furthermore, Barber still harbored plans to create a steam car syndicate. In July 1900, he had organized the Stanley Automobile Company not only as a repository for the Stanley patents but also as a means to acquire steam car patents from around the world. By May 1901, Barber was not willing to let go of the eponymous patents. Why then sell the Stanley business at a fraction of what he paid for it? The answer lies in Barber's renewed association with George Whitney.

No doubt influenced by Whitney's tales of how the Stanleys had stolen his designs for their first automobile, Barber became convinced that the Whitney patents, purchased in February 1900, not only antedated the Stanley patents but also rendered them immaterial. This conviction must have made it easy, even desirable for Barber to divest himself of the Stanley business. Soon rumors began to circulate in automotive circles that Barber planned to repudiate the brothers (Villalon 1986, 13). Gossip became fact when a Locomobile spokesman expressed the new company position in the pages of *The Horseless Age:* "Subsequent to our purchase of the Stanley patents we discovered that Whitney owned the dominating patents for steam carriages...We believe the Stanley patents are of minor value" (1902, 151).

This was a curious pronouncement indeed from the company that had launched its business on the mechanical reputation of the Stanley Brothers. Locomobile's early advertising, which boasted a picture of the twins driving in the "Stanley Locomobile Standard Carriage No. 1," had made the Stanley name synonymous with Locomobile. But now Barber was attempting to retract his company's endorsement of the Stanleys by proclaiming Whitney the originating genius of the steam motor carriage.

Meanwhile, the popularity of all steam cars in America had begun to wane. After accounting for most of the vehicles sold in the first days of the automobile business, the US steam car market was slumping by the end of 1901. Just four years after F.E. Stanley's triumphant run at Charles River Park, interest in the internal combustion engine clearly eclipsed interest in steam. Increasingly, consumers wanted gasoline cars—long before the invention of a successful self-starter and even before gas cars equaled steamers in dependability and quiet operation. According to *Scientific American,* gasoline-powered vehicles circa 1902 needed improvement in their transmission gear, sparking devices, wiring, and noise control (1903, 72). Still, in its year-end review of the state of the auto industry, the journal had to conclude: "Internal combustion engines continue to hold the undisputed lead" in consumer sales in

America (*Scientific American* 1903, 3). At the same time, the article noted that both steam and electric cars "remain competitive" in the United States.

Barber responded to these market trends by branching into production of gasoline models and retrenching, not abandoning his steam car division. For their part, the twins intended to get back into steam car production no matter what. Undeterred by the rapid shift in market preference from steam to internal combustion, and seemingly impervious to Barber's maneuvers, the Stanleys concentrated on building the best steam car they could, caring only that their product appeal to those with enough mechanical know-how to appreciate it.

The brothers had to put their production plans on hold, however, for a few months following the repurchase of their automobile business. In May 1901, Barber had not yet finished building his new central headquarters in Bridgeport, Connecticut. The brothers could not occupy their old Watertown factory until Barber moved out. According to *The Horseless Age*, Locomobile finally vacated the plant in September 1901. This meant that during the summer and early fall the Stanleys had to resort to cottage-style production, assembling a few models at a time through the agency of local carriage makers, as Flora Stanley tersely suggests in this diary entry dated August 10, 1901: "Freel began to deliver automobiles. Two went out today" (Stanley Museum Archives).

Judging by their absence from the exhibition, the brothers probably were not in full production even by the time of the Boston Automobile Show. *The Horseless Age* reported on the Boston show in its November 15, 1901, issue without any mention of the Stanley brothers. In fact the new Stanley model was not even reviewed in the pages of that journal until January 29, 1902, when it got a favorable reception:

> The new steam carriage manufactured by the Stanley Brothers of Newton, Massachusetts, embodies some novel and very interesting features. Chief among these is the new boiler... designed to combine perfect safety, high economy, and freedom from burnouts... (142).

The article went on to praise the improved technology of the new Stanley, noting that this model generated superheated steam without "the usual recourse to intensely heated coils." This Stanley model also demonstrated more efficient water use and fuel economy. It had lever steering, pedal-operated reverse, and its wheelbase was extended to six feet.

At Locomobile, management viewed with concern the arrival of this new Stanley model. A resurgent Stanley Company clearly posed a threat to Locomobile's share of the rapidly contracting steam car market (Villalon 1986, 12). In addition, the White Sewing Machine Company of Cleveland, Ohio, had successfully sold its first steam cars in 1901. In the face of growing competition for a dwindling number of customers, Barber intensified his campaign against the Stanleys and launched an all-out public relations and legal assault on the twins. He previewed his intended litigation in the January 29, 1902, issue of *The Horseless Age*. A Locomobile spokesman announced that the Whitney Motor Wagon Company sought to recover damages for patent infringement from F.E. and F.O. Stanley of Newton, Massachusetts, and to enjoin them from building their new model. Though Whitney appeared as the complainant, it was no secret who stood behind the litigation: "The Locomobile Company controls the Whitney Motor Wagon Company and is back of these proceedings" (*The Horseless Age* 1902, 151). Indeed, Barber hoped to prove in court what Whitney himself believed: that the Stanleys had copied his invention. Having declared the Stanley patents virtually worthless, Barber intended to use them to put the brothers out of the steam car business for good.

The Locomobile suit, as outlined in the pages of *The Horseless Age*, turned on three allegations: 1) The Stanley machine, introduced at the Charles River Park Velodrome in 1898, infringed upon the Whitney vehicle owned by attorney George Upham and displayed during a bicycle exhibition at the Mechanics Hall in Boston in February 1897; 2) The Stanley motor vehicle was produced only after that exhibit and closely copied the Whitney steamer in both a general and mechanical way; 3)

The new Stanley model, though somewhat altered, "possesses all the fundamental principles of the first one" (151).

To heighten the impression that Whitney and Barber were aggrieved parties, the company spokesman was less than candid when he stated the chronology of events to *The Horseless Age* reporter: "We consider the Stanley brothers have treated us very shabbily, as after selling us all their patents they followed by themselves manufacturing steam vehicles and offering them for sale at a lower price" (1902, 151).

This statement neatly eschewed salient facts: 1) The Stanleys were contractually constrained from building cars only until May 1, 1900; 2) Barber sold them back their business and their factory a whole year later in May 1901; 3) It took the brothers almost six months after repurchasing their business to get a new model into factory production.

The Horseless Age reported the Locomobile allegations with skepticism and noted that the intended litigation came in the wake of a failed effort to organize a steam vehicle trust—a probable reference to Barber's financially wobbly Stanley Automobile Company, which had been formed to purchase steam patents worldwide. *The Horseless Age* concluded that Locomobile was now pursuing a different path to domination in steam car manufacture: "(C)oming so soon after the attempt at consolidation, makes it appear that the vigorous steps are proposed to end or abate competition in the steam car line" (1902, 130-131).

Barber soon proved the journal right by filing suit not only against the Stanleys in February 1902 but also against the White Sewing Machine Company, the Foster Automobile Company, the Prescott Automobile Company, and the Grout Brothers (Villalon 1986, 16). It is hard to say how much more than chest beating some of this litigation turned out to be. As the cases began wending their way through the courts, coverage in the trade magazines all but ceased.

The Horseless Age had vowed to report every development in the Stanley suit but never mentioned it again. As the years dragged on, the news value of the great steam car litigation simply dried up. The growing consumer preference for gasoline cars no doubt contributed significantly to the lack of press pursuit. Even Locomobile seemed to lose interest in the litigation, perhaps because steam car production proved increasingly unprofitable throughout 1902. Furthermore, the Whitney patents at the center of Barber's case against the Stanleys soon became a financial liability for Locomobile.

As he had once been in arrears to the Stanleys, Barber now had trouble meeting payments to Whitney for his patents. To bail himself out of this financial strait, he turned to John B. Walker and offered him half ownership in the Whitney patents. Remarkably, Walker agreed and the Whitney patents were assigned to the Stanley Automobile Company, of which (it will be remembered) Walker was a director (Bacon 1984, 86, 99). In short order, both Walker, whose Mobile Company had never gotten into solid financial shape, and Barber were unable to meet their commitments to Whitney.

At first, Whitney accepted notes in lieu of cash from both Walker and Barber. Eventually, Barber paid off his debt with goods in trade, consisting of some thirty steam Locomobiles, for which Whitney and his partner Upham "allowed Barber $15,000 or $500 each toward reduction of his debt" (Bacon 1984, 99).

With the Whitney suit still unresolved, Amzi Barber stepped down from the presidency of Locomobile at the end of 1902, while his erstwhile partner and business punching bag, John Walker, went out of business altogether. Sam Davis now ran Locomobile, which continued to build steam cars until 1904. After that, the company manufactured gasoline autos exclusively. Development of the gas-powered Locomobile was in the hands of a new chief designer, Andrew L. Riker, who had once built electric cars (Bacon 1984, 100).

Amzi Barber died in 1909 and Sam Davis three years later, but the company stayed in business until 1930. In fact, the gasoline Locomobile became a niche market success and, according to Bacon, by 1910 "had achieved a reputation as one of the great luxury cars in the world" (Bacon 1984, 100). It was also one of the most expensive cars sold in America at that time. In 1922, the six-passenger Locomobile sedan cost $11,300. The company's manufacturing motto, "never more than four cars a day," accounted in large part for the weighty price tag. The high premium placed on quality production apparently appealed to young Raymond Stanley, who noted in his college days that he might like to own a Locomobile dealership, though he does not appear to have actively pursued the idea.

After Locomobile converted fulltime to internal combustion technology, George Whitney claimed to lose interest in building automobiles and went to work instead inventing an automatic press to improve the manufacture of asphalt paving blocks (Bacon 1984, 101). Whitney never lost his zeal or his patience, however, for proving in court that he, not the Stanley brothers, was the foremost developer of the steam car in America. His suit took four tedious years to work its way through the courts from the filing date in February 1902 until the final decree in January 1906. All the while Whitney followed the proceedings like a politician hoping for a favorable recount.

The gist of Whitney's complaint was simple: the Stanleys pirated his invention and made an illicit bundle doing so. The legal language of the suit averred that Whitney was "the original, first and sole inventor of a certain new and useful improvement in 'Motor Vehicle,'" and that the Stanleys had infringed upon this patent by copying the essential model and selling it as their own. This claim was based on U.S. patent #652,941 applied for on April 30, 1897, and granted to Whitney on July 3, 1900:

> Yet said defendants, well knowing the premises and the rights and privileges…secured by said Letters Patent, but contriving and confederating to injure your orator, and to deprive it of the profits, benefits and advantages which might, and otherwise would, have accrued to it from the invention, from and after the issuing of said Letters Patent, …as your orator is informed and believes, unlawfully, wrongfully, without license or permission of your orator…did make, use and vend…motor vehicles made in accordance with and embodying and containing said invention

of said Letters Patent...(U.S. National Archives. Circuit Court of the U.S., District of Massachusetts. *Whitney Motor Wagon v. Freelan O. Stanley, et. al.* 1570 {1902}).

Whitney sought both to enjoin the brothers from manufacturing any more cars and to make them pay him restitution for "gains and profits" accrued from the sale of their steamer. Though he no doubt hoped to prove the Stanleys were patent thieves, his case did not hinge on it. Under US patent law, if Whitney could prove "prior invention" then he would prevail even if the Stanleys had built their model with no knowledge of his own.

In their reply brief, the Stanleys denied copying Whitney's model and took the position that the technology necessary to build their first steam motor carriage was in the public domain:

> (T)hese defendants deny, upon information and belief, that said alleged invention has hitherto been in the exclusive possession of the complainant and those acting for and under it...and aver that said alleged invention has long been and now is in the possession of the public; these defendants deny, upon information and belief, that the public in general have acquiesced in the alleged exclusive rights of the complainant under and by virtue of the said Letters Patent...and aver that the public in general and manufacturers of motor vehicles in particular have never acquiesced in the said alleged exclusive right of the complainant...(*Whitney v. Stanley.* 1570)

The Stanleys were not alone in holding this position. The editors of *The Horseless Age* echoed a similar sentiment in reaction to news of the Whitney suit: "...(T)here has been a great deal of copying in steam design" and the originators of various features "(have) been unknown" (*The Horseless Age* 1902, 130). Belief that steam technology resided in the public domain may even have informed Sylvester Roper's decision not to seek a patent on any of the steam vehicles he built in the 1860s.

To prove that Whitney was neither first nor unique in his design, the Stanley brief cited forty-five steam vehicle patents that predated Whitney's in the United States, Great Britain, France, and Germany.

It also listed treatises by the Englishmen Hancock (1838) and Gordon (1836), whose experimental efforts with steam cars long preceded Whitney's. By inference, these were the sources the Stanleys used to teach themselves steam mechanics. The Stanley brief additionally named ten automobile makers (including the brothers themselves) who had "known or used" the "alleged invention or improvements, and material and substantial parts of the same, described and claimed as new in said Letters Patent, No. 652,941" before Whitney built his own machine.

The Whitney suit languished four years in court, and when the final decree came down on January 1, 1906, it proved a decided anticlimax:

> No deposit having been made to secure the clerk's fees in the above cause, as provided by the provisions of the Order of the Court entered October 5, 1905, it is ordered, adjudged and decreed by the Court that the bill of complaint in this cause be...dismissed without prejudice...(*Whitney v. Stanley.* 1570).

Whitney's claim had been dismissed on a technicality; the judge never got to rule on the issue of patent infringement. Within a week of the judge's ruling, however, Locomobile and Whitney filed a second suit. But in this next legal round, the scope of the alleged infringement had shrunk dramatically. Whitney and Locomobile did not again contend that the Stanleys had copied an entire automobile but declared infringement upon Whitney as the "original and first inventor of certain new and useful improvements in Hydro-Carbon Burners" (U.S. National Archives. Circuit Court of the U.S., District of Massachusetts. *Whitney Motor Wagon Co. v. Stanley Motor Carriage Co., Freelan O. Stanley, Francis E. Stanley* 214 {1906}).

In a letter to the editor of *Motor Age*, Frank Stanley discussed the background of this second suit. According to Stanley, Locomobile informed the brothers sometime in 1904 that they were violating Whitney patent #777,578 held on a hydrocarbon burner. The brothers tendered substantially the same answer to this second allegation as they had to the first. As Frank put it, "(T)his patent involved principles which all of the manufacturers of steam carriages were infringing" (*Motor Age* 1907, 27). Stanley wrote that he and Sam Davis had dickered unfruitfully for two years over the matter. Their failure to reach an agreement resulted in the second round of Whitney v. Stanley.

On September 3, 1907, Judge Lowell of the Circuit Court of the United States, District of Massachusetts, rendered his final decree: "...(U)pon pleadings and stipulation of the parties by their counsel, it is ordered, adjudged and decreed that the said bill of complaint herein be... dismissed without cost to either party" (*Whitney Motor Wagon v. Stanley Motor Carriage.* 214 {1906}).

Simply put, Locomobile and the Stanleys had finally succeeded in settling their case out of court. According to Frank Stanley, the agreement resulted in the brothers purchasing "all of the patents which the Locomobile company had obtained from all sources" (*Motor Age* 1907, 27). This meant the Stanleys not only got back all their own patents but also owned as many as seventy patents from other steam car makers. After many years, the brothers again controlled their own destiny, free from any future court action. For its part, Locomobile got compensation for patents it no longer needed. But for Whitney, the emotionally freighted issue of inventor's precedence remained untouched by the settlement.

By the time the second suit was filed, the allegation that the Stanleys had stolen a whole automobile design from Whitney had frankly run out of gas. In 1903, the Stanleys had re-engineered their steamer so that it no longer significantly resembled Whitney's design. Perhaps in a backhanded way, the twins owed Whitney a thank-you. The threat posed by his lawsuit set off an inventive flurry, at least for Frank. Freel's declining health at this time probably meant that he had little to do with the drafting.

Up to this point, steam cars, including Whitney's, used a simple bicycle ratchet chain to drive the rear axle. What Stanley did was mount the engine horizontally on the running gear so that it drove the differential on the rear axle through a spur gear and pinion.

Scientific American heralded this innovation as the "latest practice in steam cars...inaugurated by the Stanley Brothers, the original inventors of the light steam car in America" (1903, 286).

Poor Whitney, his contributions to automotive science seemed forgotten less than ten years after his initial success. Furthermore, the chances of reconstructing his reputation and winning restitution in court were all but dashed by F.E.'s new design. On June 9, 1903, Frank Stanley won a patent on a steam vehicle with engine on the axle and housing. That patent grant no doubt convinced Locomobile of the futility of ever refiling the charge that the Stanleys were building their cars from a stolen design.

Throughout his long lifetime, Whitney clung to his conviction that the Stanleys had outstripped him in automobile manufacturing because they had pirated his invention and with it, his rightful place in automotive history. In time, he looked back on his two court suits against the twins as complete victories, even though one case was dismissed and the other settled. John Bacon, author of *American Steam Car Pioneers*, apparently shared and fostered this idiosyncratic version of events in his steam car history. Bacon wrote that Whitney "subsequently won... both infringement suits," with the result that "the Stanleys had to redesign their car." (1984, 63).

Although there was no legal imperative for the Stanleys to change their automobile design, there certainly was incentive. Ironically, the court battle with Whitney proved to be a boon to the Stanleys' reconstituted automobile business. The new Stanley models proved so popular with steam customers that by September 1, 1903, the brothers already had sold six hundred of them. At least that is the figure F.E. quoted George Eastman when they met for lunch in Boston to renew negotiations for the sale of the Stanley Dry Plate Company.

The Stanley Motor Carriage Company at 44 Hunt Street.
Watertown Free Public Library.

CHAPTER 11

King of the Road

ALMOST A YEAR after the Stanleys had begun to manufacture steam cars under their own name again, they still derived the bulk of their income from the sale of dry plates. Indeed, profits from the Stanley Dry Plate Company made possible the building of Frank's first car, the partnership with Walker and Barber, the repurchase of the factory, and the redesign of the chain drive. But when Frank met Eastman in November 1903, it was clear the twins were ready to enter the car business fulltime. F.E. no longer cared whether Eastman purchased the dry plate factory along with the company name. In fact, as Eastman wrote to C.S. Abbott in a letter dated November 27, 1903, Stanley was decidedly eager to hold onto the property:

> I got F.E. Stanley on the phone and asked him if he had found any use for that real estate yet. It took him a few seconds to grasp the idea but when he did he seemed quite enthusiastic. He said yes, they had bought a lot more property adjoining the dry plate works and were putting up a new automobile factory…"(George Eastman House Archives).

Sterling Elliott's Old Hickory Bicycle factory, which the twins had bought back from Amzi Barber, could not meet the growing volume of automobile sales. So the brothers purchased a tract of land next door and planned to incorporate the bicycle works and the dry plate factory into one large plant.

Construction of the new automobile headquarters had already begun by the time F.E. met Eastman for lunch to resume sales negotiations. On October 12, 1903, Watertown, Massachusetts, issued the Stanleys a permit to put up a four-story brick structure at 44 Hunt Street. Two weeks after obtaining the permit, however, a dispute with the bricklayers' union had the brothers petitioning for a change. They sought to substitute "Portland cement concrete" for brick. The substitution was granted and as a result the Stanleys, with the H. F. Ross Company as architects, erected one of the earliest reinforced concrete factories in the United States. Their plant was not the first or the tallest—the latter distinction was held by the 210-foot Ingalls Office Building in Cincinnati—but it did help break new construction ground (www.enr.com.2/1/99). The original building stood fifty feet high, fifty feet deep, with a frontage of 162 feet. The shell of the building was poured concrete; the floors were wood frame. The new factory cost the Stanleys about $35,000 to build.

When the brothers started assembling cars at their new facility, they still marketed them under the Stanley Dry Plate name. Not until the last glass plate was sold to George Eastman in January 1904, did the fledgling automobile company finally get its own name. The Stanley Motor Carriage Company (SMCC) was incorporated in Massachusetts on September 19, 1904, with a capital investment of $95,000.

The launching of their second corporation occasioned no change in marketing philosophy or strategy for the Stanleys. The brothers were still convinced the superiority of their product would sell itself without the agency of modern advertising. Once again the Stanleys intended to take their wares to the American consumer. But it was no longer necessary to barnstorm the nation just to pique people's interest in the motor car. In the five years since F.O. first traveled the countryside assembling his Locomobile and offering demonstration rides to farmers, the American automobile industry had lost much of its ambivalent novelty. The annual auto show had become a fixture in the nation's largest cities. Despite a dearth of racing venues and the generally poor state of the nation's roadways, automobile promoters and enthusiasts had begun to hold a number of well-attended annual events—auto parades, hill climbs, and speed races—where manufacturers and buffs alike could show off their machines.

One of the earliest auto rallies in America was launched long before the organization of the Stanley Motor Carriage Company. The first Vanderbilt Parade of Automobiles took place in Newport, Rhode Island, in 1899. William K. Vanderbilt Jr. himself took first prize for "speed and maneuvering skills" in his Stanley-designed Locomobile (*The Horseless Age* 1899, 6). But the rich and generally staid residents of this seaside community proved inhospitable to the motoring enthusiasm of Cornelius Vanderbilt's great grandson. His "wild" driving over public streets apparently occasioned the imposition of Newport's first speed limit of six miles per hour.

The wealthy automobilist soon moved away to Long Island, where he began planning for an American race that he intended to rival the French Grand Prix. His new neighbors were less conservative than the crowd at Newport, but Vanderbilt's plan to set a racecourse over local roads drew vigorous objections. A few newspapers railed against the speed maniacs, and a group of farmers, claiming the race would impede their ability to get produce to market, went to court to stop it. But Vanderbilt prevailed, and on October 8, 1904, under the sponsorship of the American Automobile Association, the first Vanderbilt Cup Race took place. The voices of opposition did not keep the public from coming. On the day before the event, the Long Island Railroad ran special trains packed to the aisles with spectators (Adcock, www. lihistory.com).

Early in the chilly fall morning, seventeen competitors set off on the thirty-mile-long triangular course that was laid out over three turnpikes in Nassau County. The racers had to cover the distance three times. An American named George Heath led the field, but his victory was tinged with disappointment for Vanderbilt who wanted an American car as well as an American driver to take the prize. Heath had driven to victory in a ninety-horsepower French-built Panhard. Vanderbilt hoped that a race

F.E. Stanley and driver Fred Marriott pose with one of the Vanderbilt racers. Though they built it without a condenser, the Stanleys intended this car to compete at long distances, but its greatest successes came at hill climbing events. Fred Marriott Collection.

Speed Age, *Stanley Museum Archives.*

on US soil would spur American manufacturers to produce a car that could beat the big European models.

Except for Locomobile, no American automobile maker ever posted much success at the Vanderbilt Cup competition. The Stanleys built two long-distance racers expressly for the event but these cars never competed in Long Island. They had been purchased in fact for the race in 1906 by two members of the Cape May Automobile Club, Charles Swain and John N. Wilkins. During the months preceding the race, F.O. Stanley assured the men that their automobiles would be delivered in time for the qualifying heats in August. But ill health took F.O. abruptly to Denver at the same time that mechanical difficulties surfaced with the cars. In an unapologetic letter to his clients, F.O. explained that the racers would not be fit to go after all because neither he nor his brother was available to supervise the repairs. Swain and Wilkins, while expressing sympathy for F.O.'s health, were furious at the cavalier manner in which the SMCC had dropped their order. In an outraged letter to the editor of

Automobile Magazine, they declared, "(we were) ignominiously thrown down at the last moment" (1906, 297).

Publicly, the Stanleys never admitted to having mechanical difficulties with the racers or to abandoning their clients but alleged that race officials had banned steam entrants on the grounds that their foggy emissions posed a hazard to other drivers. This story, which was less than candid, provided a credible screen for the brothers' serious public relations blunder because it came at a time when steam car racers (as we shall soon see) were wearing out their welcome at other venues.

As it turned out, the Vanderbilt race in 1906 proved memorable for more than the Stanleys' absence. One man was killed and several other spectators injured after they wandered onto the narrow course, presumably to get a better look at the oncoming racers. The tragedy prompted Vanderbilt to begin building a roadway exclusively for automobiles. Two years later, nine miles of the Long Island Parkway

Following his outstanding (but not winning) run up Mt. Washington, Frank Stanley maneuvers his Model B runabout down wooden planks over the steps to the platform at the Summit Lodge.

Courtesy Erik Haartz, Stanley Museum Archives.

had been completed in time for the running of the Cup and were incorporated into the racing course. Although Vanderbilt employed the most advanced highway design of his time—limited access and overhead passes—his forty-five-mile-long reinforced-concrete pike proved a financial flop. Nor could Vanderbilt convince the State of New York to buy it from him. He fared no better with the fate of his Long Island race. Despite Vanderbilt's best efforts, the venue never became popular with European drivers. The Automobile Club of France put it with Gallic terseness in the pages of *Automobile Magazine*: "We ignore the existence of the Vanderbilt Cup Race" (1906, 943).

The Vanderbilt's dream might have dwindled into oblivion but for the race of 1910. The carnage that year included four spectators killed and twenty injured. Tragedy brought racing on Long Island notoriously to an end. Thereafter the center of automobile road racing in the United States moved to the newly opened speedway in Indianapolis.

Lighter in spirit than high-stakes speed racing, hill climbing was a popular event among a wide variety of motorists. From the time Frank Stanley bumped the bar at the top of the ramp at Charles River Park, driving an automobile up a steep incline became a staple of automotive competition. In 1905, the Automobile Club of Worcester, Massachusetts, invited motorists to take on the city's famed Dead Horse Hill. Just seven years earlier, most drivers had failed to negotiate the ramp at Charles River Park. Due to the rapid progress in automotive technology, the Worcester Climb was deemed a "jest" for the entrants the first year it was held. In fact, automobile hill climbing was routine enough by this time to attract (and permit) women drivers to compete.

At the Worcester competition in 1906, Mrs. H. Ernest Rogers (née Margaret Dixon) of Brookline, Massachusetts, won three gasoline-car events in her ten horsepower Maxwell. But H.F. Grainger captured the over-all championship that year in a Stanley car with the winning time of one minute thirty-one seconds. Fred Marriott, who would become the most famous Stanley driver of all, drew duty that day behind the wheel of one of the yet-untested Vanderbilt racers. Marriott's effort revealed the

Steam on steam: F.E. Stanley and Joe Crowell betray nothing of their eight-mile jolting ride to the top of Mt. Washington as they sit stoically for a picture with an engine from the Mt. Washington cog railway. Photograph by Walter R. Merryman.

Courtesy John Burnham, Stanley Museum Archives.

mechanical flaws that knocked the specially built automobiles out of the Vanderbilt Cup (*Automobile Magazine* 1906, 857-858).

When the Stanley racers finally proved roadworthy, one of them scored a triumph not in Long Island but at the Worcester Climb of 1908. Driven by L.F.N. Baldwin, the racer set a new course record of 55.5 seconds.

The Horseless Age noted that this Stanley victory "once more proved the superior climbing ability of the steam car" (1908, 694). The next year, in the same Stanley model, Baldwin bettered his own mark with a time of fifty-four seconds flat, which stood as the permanent course record (Hart 1985, 9). In fact Stanley steam cars so dominated hill-climbing events that when H. L. Bowden of Waltham, Massachusetts, challenged F.O. Stanley to a head-to-head competition between Bowden's eight cylinder Mercedes and any Stanley car for $1000, F.O. reputedly agreed, provided that the Mercedes race from the top of the hill down. A disgusted Bowden withdrew his offer (H. Elliott 1945, 35).

If hill climbing in places like Worcester was fast becoming routine, the "Climb to the Clouds" race up Mt. Washington was anything but tame and anything but a hill! First staged on July 11 and 12, 1904, the Climb to the Clouds had contestants retracing F.O. Stanley's historic run up the precipitous peak but this time against the clock. Freel's health did not permit him to attempt to reprise his triumphant ascent of a few years earlier, so Frank Stanley upheld the family honor this time. As the racing date approached, Stanley and other wealthy motoring sports, primarily from Boston and New York, converged on the White Mountains. This was what the owners of the newly opened Mount Washington Hotel and its sister hostelry on the other side of the mountain, the Mount Pleasant, had hoped for when they sponsored the climb. On the eve of the race, a party atmosphere pervaded both establishments as competitors and their entourages awaited the climb (*Stanley Museum Quarterly* [1904] 1994, 11).

The Climb to the Clouds consisted of a series of heats run over a two-day period. One at a time, competitors raced the clock and their nerves to the top of Mount Washington via the same road Freel and Flora had traveled in 1899. Members of the board of the American Automobile Club served as officials, and the Chronograph Club of Boston kept the times. Telephone lines connected the start and finish lines, with three relaying points in between. Thus the racers stayed apprised of their split times, and the officials knew just when to send off the next car.

On the first day of competition, Frank Stanley drove his small red six-horsepower 1903 runabout to the top in just over half an hour (31:41 2/5). Luckily, he did not have to stop to jack up the rear wheels because Stanley autos now had greatly improved capacity to pump water to the boiler.

Before the race, F.E had made bold predictions about the performance of his Stanley. On day two of the race, he set the steam record for the course, bettering his time to 28:19 2/5. Having made good on those claims, he was greeted with admiring enthusiasm by onlookers at the Summit Lodge. As a bit of a publicity stunt, Stanley agreed to drive his car up wooden planks to rest on the platform of the Summit for a photo session—steam on steam—with a locomotive from the Mount Washington Cog Railroad, which had its terminus at the lodge. At least the hoopla gave him and his young "mechanician" Joe Crowell, who had ridden along stopwatch in hand, a chance to shake off the effects of their eight miles of jolting. It is painful to contemplate what sort of grueling ride the two endured, since the little steam car had been stripped literally to its skivvies, with even the seat cushions removed. Crowell, it will be remembered, had made his first appearance on Mt. Washington the year after Flora and Freel's ascent. With his mother along for ballast, he managed to best F.O.'s time to the top by six minutes, thanks to the first hand-held water pump on a Stanley car.

Artist Peter Helck captured the excitement of William Hilliard's' triumphant finish in the 1905 Climb to the Clouds race. An expenditure of over $18,000 and 60 horses under the hood carried him to victory. Painting by Peter Helck.
Stanley Museum Archives.

Stanley's was the time to beat on that second day of racing, and Harry Harkness, a New York millionaire, proved the man to do it. F.E. ultimately had to settle for second place behind Harkness' $18,000, 2200-pound sixty-horsepower Mercedes. Averaging twenty miles per hour over the rocky and rutted course, Harkness risked his life to get to the top in just twenty-four minutes, thirty-seven and three-fifths seconds. A reporter from the *Lewiston Evening Journal* chronicled the breathtaking highlights in a story that ran on July 18, 1904: "He [Harkness] often skidded dangerously near the edge...knock(ing) huge stones down into the valleys of the mountains....(S)o fast did he take the hummocks and water ruts that he was more frequently out of the saddle than in..." The two hundred or so intrepid fans who had climbed up or ridden the rack railway to the top greeted Harkness "with a mighty cheer"(*Stanley Museum Newsletter* [1904] 1991, 16).

Although the mountain ascent was clearly the show-stopping event in the Climb to the Clouds of 1904, after the vertical competition was finished, contestants were invited to participate in a ninety-mile road tour that took drivers in a circuit around Mount Washington. Participants were to set out from the Mount Washington Hotel in Breton Woods and finish at the Profile House near Franconia Notch. Charles Jasper Glidden, telephone pioneer and later aviator, organized the event under the sponsorship of the American Automobile Association. Glidden had already established himself as an inveterate automobile trekker, having traveled successfully to the Arctic Circle and back in 1901. The following year he and his wife logged 45,000 miles touring through twenty-nine countries in a British-built Napier. As routine equipment, the Gliddens carried with them a set of flanged wheels they could affix to their car in case railroad tracks proved to be the only pathways through some of the lands of their travel (Delong, www.vmcca.org/bh/cjg.html).

In April 1904, to help kick off the St. Louis World's Fair, Glidden took part in an AAA sponsored auto rally from New York to Missouri. Buoyed by the experience, he came away convinced that long road expeditions were the best way to demonstrate both the automobile's

In 1905 F.E. Stanley again challenged the precipitous course, this time in a Model H speedster. But a broken water indicator and a burned out fusible plug dashed Stanley's hopes for victory. Photograph by Walter R. Merryman.

Stanley Museum Archives.

fitness for distance travel and the pressing need for major renovation of the nation's roadways. Glidden's satisfying experience with the World's Fair road rally gave him the idea for the Mount Washington auto tour. In turn, the success of this event launched the longer, more-famous road-a-thon he sponsored for eight years from 1905 to 1913.

In the Mt. Washington mini-tour of 1904, F.E. Stanley entered the rally with a factory-fresh black 1904CX. He was accompanied for this journey not by his mechanic Joe Crowell but by his wife, Gustie, who in middle age was a woman of considerable girth and no small fear of automobiles. Over the years, she confided to her diary great trepidation about speed and cars, but this once, perhaps not to be outdone by the memory of Flora Stanley's brave jaunt up Mt. Washington, Gustie conquered her racing fears.

She was dressed, according to the fashion of the day, in a motoring costume that included a pongee silk duster and a large veiled hat (in lieu of goggles) tied under the chin with a scarf. Her highly visible chapeau turned out to add immeasurably to the delight of the spectators. She and Frank started the tour in seventh position, and as the *Lewiston Evening Journal* reported it, every time F.E. got ready to pass a competitor there would be a brief conference, followed by Gustie grabbing her motoring hat as the Stanley whizzed ahead. With each grab for the hat, she telegraphed Frank's next passing maneuver until he took over the lead from Glidden himself riding in the "fastest and most powerful gasoline car in the procession." The Stanleys finished in time for Gustie to compose herself to greet the rest of the field from the porch of the Profile House.

In an account of the day's events, the *Lewiston Evening Journal*, enthusiastic over the exploits of its famous former resident, dubbed Stanley's little car "King of the Road." According to F.E.'s account, it was Glidden himself who informally bestowed that title when the Stanleys passed him on the tour. As recorded by the *Lewiston Journal*, Glidden said: "Mr. Stanley, you have a marvelous machine. You can beat us up hill or on the level. You are king of the road of this contest"(*Stanley Museum Quarterly* 1991, 15). After acknowledging the CX's superior performance, Glidden reputedly asked that the "king" not pass his car "for the rest of our journey."

Later in August, Frank and Gustie teamed up again to motor from their home in Newton to Kingfield, Maine—a journey of 236 miles—in a single day. No blown tires, no broken pumps, "not even so much as a loosened screw" troubled their journey. On this day the steamer performed with matchless ease. Mrs. Stanley, clearly basking in the newspaper's characterization of her as a fearless motorist, recorded the event in a letter dated August 26, 1904, to the *Lewiston Journal*: "We flew by the cities and towns so fast," she wrote, "that it almost seemed as though we were standing still and a panoramic view of the surrounding country was being presented to our view" (*Stanley Museum Newsletter* 1991, 3).

The summer of 1904 proved to be an important one for the newly launched SMCC. Frank's victory over the larger, more-powerful gasoline automobiles in the Glidden mini tour served convincing notice that the Stanleys were not ready to concede automotive performance to the internal combustion engine, even if steam had already lost out in sales. Two years later, racing events on the extraordinary sand beach at Ormond-Daytona in Florida would prove that the Stanley victory at Mt. Washington had not been a fluke. In January 1906, the name Stanley would pair indelibly with speed and fix into legend the image of a steam automobile, throttle open, moving faster over land than any other thing on wheels.

CHAPTER 12

The Fastest Thing on Wheels

THE REASON THAT A SIX-HORSEPOWER Stanley could administer a drubbing to the early gas giants lies in the nature of steam cars themselves. Once the pressure is built up, steam engines are fully powered from rest and thus generate massive torque or force for rotation. Maximum power is available at any speed from zero to five or fifty miles per hour: steamers have the advantage of direct drive; they need no clutch or transmission. This is what makes them formidable hill climbers and superb racers: "If the steam pressure is high enough to slip the wheels, the steam car has the greatest acceleration that is possible in an automobile" (Derr 1932, 9).

A steam car, even one with considerably less horsepower than its gasoline rivals, had a great advantage in the "rolling start" type of race that was popular in the early days of automotive competitions. The driver, having warmed up earlier, simply wheeled slowly to the start line with the throttle down, while the gasoline competitors had to creep along in their lowest gears. When the starter gun fired, the steamer shot forward as soon as the driver opened up the throttle. The gasoline car had to shift through gears to accelerate, thereby losing competitive seconds in the process.

During the first decade of the twentieth century, few American racecars—steam or gasoline-powered—had much chance to test their mettle at home. All significant auto races took place in Europe, and all important speed records belonged to Europeans. This resulted from the good condition of roads abroad and, in no small measure, from the deplorable condition of the roads at home, which had improved only slightly since F.E. Stanley remarked upon their inferiority in the summer of 1899. American automobilists, nevertheless, were ever on the lookout for suitable racing venues that might lure international competitors to these shores.

In the early 1900s, several automobile enthusiasts began to tout the thirty-mile stretch of beach between Ormond and Daytona, Florida, as a favorable site for auto racing. The sand on this smooth beach contains about ninety-seven per cent quartz, making the surface at low tide remarkably hard. Winter guests at the Hotel Ormond had long enjoyed the special properties of this beach in leisurely biking or the more adventurous activity of "bicycle sailing." With a sail rigged to the frame of a bike and the luck of a favoring wind, it was possible to run seventeen miles along the beach without ever peddling. Even a lady's parasol sufficed as a sail when a stiff breeze was running. More recently, the sport of sand or tricycle sailing, which worked on the same principle as iceboat sailing, was gaining popularity on the beach (Punnett 1997, 3, 13).

The first person to call national attention to the fine conditions at Ormond-Daytona was a winter vacationer named C.W. Birchwood. An early automobilist himself, Birchwood discovered in 1902 that running a car over that smooth beach was as effortless as sailing. He wrote an article detailing the racing properties of the beach for *Automobile Magazine*. Birchwood's "find" was soon tested by J.F. Hathaway, a well-to-do former businessman from Massachusetts who actually marked off a mile-long course on the beach and managed to drive his steam runabout faster there than he had done earlier on a course in Rhode Island (Punnett, 1997, 4, 6). Now a believer, Hathaway immediately launched a concerted effort to win backing for an automobile-racing event at Ormond-Daytona. He wrote enthusiastically in *Automobile Magazine* that the beach was some 400 to 600 feet wide at low tide and the sand surface as hard as asphalt. An automobile weighing 5000 pounds, he asserted, "makes no impression on the hard sand." Hathaway prophesied that "world's records will be made (here) in the future" (1903, 181).

Picking up the drumbeat, *Automobile Magazine* sent one of its staff writers, William J. Morgan—a man with considerable public relations experience—to check out the Florida conditions firsthand. Morgan quickly became as enthusiastic as Birchwood and Hathaway. With the endorsement of his magazine and the financial backing of the managers of the Hotel Ormond, who coincidentally ran the Mt. Washington and the Mt. Pleasant hotels in New Hampshire, Morgan managed to schedule a late winter race in March 1903. The expectations of the Ormond proprietors were met immediately as *Automobile Magazine* reported that the prospect of the races "has delayed the late season departure of Southern visitors" and "crowds are beginning to arrive" at Ormond and Daytona hotels (1903, 363). A good deal of infighting between rival automobile associations in Florida resulted in a small field of competitors for the first year's events, but among those who did show up was gasoline auto maker, Ransom E. Olds. The following year Henry Ford entered his "Arrow" racer with its new Model B type engine (Punnett 1997, 18). Having set a flying-mile record (91.37 mph) at the wheel of his racer on a course plowed across the frozen surface of Lake St. Clair, Ford was looking for another chance to demonstrate the power of his latest engine (http://media.ford.com/article_display.cfm?article_id=7242).

The Stanleys first participated in the Florida races in January 1906, after a fellow Newtonite and steam car maker named Louis Ross made racing history at the event in 1905. Ross had competed at Ormond-Daytona during the previous year with a 1904 six-horsepower Stanley that he drove to a new world steam car record for the mile (55 2/5 seconds). William K. Vanderbilt, however, had dominated all events that year in his ninety-horsepower Mercedes, winning among many honors, the Dewar's cup mile race. The latter consisted of three rolling start heats followed by a deciding fourth heat among the winners. In 1904, Ross had prevailed only in the steam division. But 1905 would be a different matter.

Ross returned to Florida with a racing car of his own design and construction. His new racer, quickly nicknamed the "Wogglebug" for its tendency to yaw, had a low-slung, sleek torpedo shape and was powered by two separate Stanley engines, each mounted to drive one rear wheel (Punnett 1997, 32). (The difficulty of steadily feeding power to two separate wheels may have caused the car to "woggle" from side to side on occasion. Ross's most celebrated incident of weaving on the track, however, occurred at Narragansett Park after blowing a tire). His woggling notwithstanding, Ross had already tested the mettle of his racer on the Readville track in Boston in June 1904 when he broke the steam car record for five miles. He fared even better that winter at Daytona, as he blew by the high-powered European-built competition (including Vanderbilt's winner from the previous year) to capture the Thomas Dewar trophy.

The international attention stirred by Ross' steam-powered victory was not lost on F.E. Stanley, who had recently tasted a bit of his own racing success at Mt. Washington. Furthermore, if Stanley engines were to power a vehicle to a world title, they might just as well be propelling a Stanley-made racer. This doubtless was the argument that family friend and Ormond enthusiast J.F. Hathaway brought to the SMCC in hopes of convincing Frank to enter the Florida race. Hathaway may also have played to F.E.'s love of speed and his well-known desire to demonstrate the superiority of his automobiles. Whatever incentive he used, one imagines Hathaway didn't have to work too hard to make a sale. Over the concerns of Gustie, F.E. made up his mind to build a racer in time for the Ormond events the following January.

Since poor health was still keeping F.O. on the sidelines of the business, Frank had to design the 1906 Ormond entry himself. He adopted Louis Ross' revolutionary torpedo shape, which he himself may have had a hand in designing since he helped young Ross construct his "wogglebug." For the chassis, Frank used an inverted canoe made of cedar strips covered with canvas. According to Raymond Stanley, F.E. tested the wind-drag properties of several different canoe bodies by towing them through the streets of Newton with a spring scale (Punnett 1997, 62). He opted to have the Robertson Canoe Factory of Auburndale, Massachusetts, build him a special canoe body.

The Stanley racing car was designed so that the driver sat directly on the floor and steered by means of a bar much like a boat's tiller. The red racer was equipped with a two-cylinder thirty-horsepower steam engine mounted horizontally behind the rear axle. In contrast to Ross's "wogglebug," one engine drove both rear wheels. Just in front of the rear axle sat a thirty-inch boiler—complete with familiar reinforcing tensile wire. As Dick Punnett noted in his *Racing on the Rim*, this Stanley rear-engine mount predated "modern Grand Prix racers by many decades." The drawback to the design, as he further pointed out, was that the engine and boiler mounted in the rear made the Stanley decidedly heavier in back than in front.

Before the start of the mile trials competition Fred Marriott checks out the "wog."

Stanley Museum Archives.

At least Augusta Stanley could breath easily about one aspect of the Ormond event: Frank would not be at the helm of the racer. For that honor and task, he tapped the young man who headed the SMCC service department. Fred Marriott had been a bicycle racer and mechanic who, like so many other cyclists, made an eager transition to automobiles in the late 1890s. Mustachioed and rakish like the barnstorming pilots of a later generation, he proved himself to be fearless on the sand course.

Frank and Gustie traveled to the Florida races for that first season in grand style, occupying a stateroom and two drawing cars on the train. Since a number of elegant evening affairs attended the daytime events, Mrs. Stanley noted in her diary on January 23 that she had brought along her "best dress and diamond necklace." The Stanley racer, its driver and mechanics, were also transported by rail.

By January 16, the Newton entourage had settled into serious preparation for racing. The "wog," as Gustie affectionately called the Stanley racer, had operated smoothly in its Newton test-drives, but running it on the beach in Florida revealed some roughness in the engine at certain speeds. Fred Marriott described the problem in an interview with Thomas C. Marshall:

> I couldn't hold onto the wheel, (sic) my legs would get such a tingling they'd go completely to sleep. I wore heavy padding in the seat of my pants, up my back, and rubber pads on my soles helped some. We taped my hands where they touched the wheel and up to above my wrists (*Stanley Museum Quarterly* [1982] 1996, 23).

Though Stanley and Marriott figured out what the problem was, they could not fix it until they got back to the factory in Watertown. Marriott, therefore, had to endure the tingling in order to compete. His first official trial run of the mile was highly impressive (30 3/5 seconds). According to Gustie's diary, Frank was "in transports of delight" over the car's performance, which had been "better than he thought." Day two of the prerace activities, however, brought a major setback. The engine on the racer cracked a cylinder, and Frank had to wire his son-in-law Prescott Warren (now in the family business) to ship a new engine.

Meanwhile as preparations went forward for race day, Frank and Gustie had another sort of adventure. They had driven to the beach to watch glider pilot Charles Hamilton—on hand as part of the general festivities—attempt to get airborne. To the delight of the townspeople of Ormond, Hamilton proved willing to risk his neck several times in a rig that resembled a large silk and bamboo box kite. Hamilton took off from the beach as he was towed behind an automobile. Once in flight he had only a skimpy rope platform, situated between the top and bottom wings, to stand on. He steered the thing solely by shifting his weight in a kind a gyrating dance (Punnett 1990, 3, 4). On the day Gustie and Frank were spectators, Hamilton managed to get about 300 feet off the ground before plummeting precipitously to earth in a trajectory headed straight at the Stanleys in their car. A flagpole broke the glider's fall, saving the pilot and the Stanleys from serious injury. Gustie and Frank escaped unscathed and Hamilton walked away that day with only a few scrapes.

Even without the added sideshow of glider flights, excitement for the 1906 auto races was running high, bringing with it large crowds to the beaches. *Automobile Magazine* reported that many hundreds of spectators literally poured out of New York City's two automobile shows and onto special trains bound for Daytona (1906, 265). The international field of competition that "Senator" Morgan had managed to assemble that year sparked a good deal of interest in the event. The Daimler and Mercedes companies posted entries. Henry Ford returned with the six-cylinder racer that had washed out of the competition the previous year. Louis Chevrolet, then a professional racer, drove for J. Walter Christie. The French A. Darracq & Cie Automobile Company entered four cars; and Vincenzo Lancia of Italy drove a 110-horsepower Fiat. Providing inadvertent comic relief, Alfred G. Vanderbilt introduced an experimental 250-horsepower automobile that had been built expressly to set a mile record. Overbuilt might be a more apt description:

the eight-cylinder engine proved too powerful for its clutch. It took four men and a rope to turn the start crank and with eight open exhausts the car emitted a fearsome number of decibels. Fortunately for all involved, the mechanical bruiser never made it onto the raceway (Punnett 1997, 45-46, 57).

When January 23, the first day of competition arrived, the race for the Dewar's cup did not unfold quite as Morgan had hoped. Several scratches and a "no show" somewhat reduced the field. The tempestuous lead driver for Darracq, Victor Hemery, withdrew his company's 200-horsepower racer after another of the team's autos was deemed over the weight limit. Alfred Vanderbilt's monster car had too many mechanical problems to make the race, and Henry Ford's driver, suffering a serious miscue, did not arrive in time to start (Punnett 1997, 46).

The day had dawned rainy and dark but not so wet as to interfere with the quality of the course. In the Stanley camp, all was in readiness. The replacement engine had arrived in time and was duly installed. Marriott was set to go against the remaining field. As before, the Dewar's cup began from a rolling start and Marriott had the same advantage of acceleration in his steam racer that Louis Ross enjoyed the year before. In the first heat Marriott covered the course in just 32 1/5 seconds, knocking three-fifths of a second off the previous world record and deciding the race with that time. All previous world records in the mile had been set from a running start, that is, the autos were in full acceleration as they passed the starting line. It was "a great Stanley day" Gustie wrote in her diary, and that night she brought out her diamonds and best dress for the celebration dinner.

The next morning, however, Fred Marriott was back at work on the racecourse. The five-mile open championship was scheduled for day two of the events. Vincenzo Lancia won the first heat in his Fiat with a time just under three minutes (2:54 3/5). Fred Marriott was to drive in the second heat against another Fiat entry and Victor Hemery, who had re-entered the competition after the fourth Darracq racer finally met the weight limitation. A false start called the racers back to the line, but the Frenchman refused to comply and was disqualified from the competition. With a fair start, Marriott outdistanced the field in a world record time of 2:47 3/5. During the run, a casting cracked on the Stanley and Frank wanted to withdraw the racer from the final heat. The race officials, no doubt thinking of the world attention that was now focused on events at Ormond, denied the request for a scratch, thus setting up a third heat showdown with Lancia. The drama fizzled about halfway into the race, when Marriott had to shut down his boiler causing the Stanley to slow to a third place finish.

The crew of Stanley mechanics was nevertheless able to get the racer back in shape for the next day's events and in time for Marriott's finest hour. In the flying start measured mile he not only trounced the competition but also demolished his own world mark set in the Dewar's cup two days earlier. Marriott broke the "30 second barrier" in the time of 28 1/5 seconds, which clocks out to be 127.659 miles per hour. Louis Chevrolet made a good run at Marriott's mark, posting a mile in 29 2/5 seconds (*Automobile Magazine* 1906, 274). But it would take four more years before Barney Oldfield knocked .87 seconds off the Stanley's record time in a big Benz gasoline car.

Marriott takes a test run on the beach.

Stanley Museum Archives.

Doubtless, this January day belonged to Fred Marriott. As the *Florida Times Union* of Jacksonville saw the events of January 26, it was a patriotic triumph: "He (Marriott) has beaten the boasted cars of Europe. He has outdone the premier drivers of the Old World. He has defeated time itself, annihilated distance, and the chief glory of it all is that he is an American and drove an American built car."

From Denver, brother Freel telegraphed, "Wonderful, wonderful, wonderful" (Stanley Museum Archives). According to the *Portland Advertiser,* February 1, 1906, edition, F.E. graciously credited Fred Marriott with the Stanley triumph: "I wish to give every possible credit to Fred...I know it took courage and a good head to guide the steamer as it flew over the beach at the highest speed anything ever traveled on wheels."

The next day, Marriott won the thirty-mile race for American-built cars in a time of 34:18 2/5. Frank Durbin, in a 20-horsepower steamer, added to the Stanley total with a victory in the fifteen-mile price handicap for American touring cars (*Automobile Magazine*, 1906, 274). But no Stanley driver entered the grand finale one-hundred-mile race because that distance was beyond the range of a steam car without a condenser, and no Stanley-built model at this time had one, although the technology existed.

Marriott was, however, very much in contention for the two-mile record on the very last day of the Ormond-Daytona events. Considering how the "thirty second" mile had been demolished, race officials and spectators alike had reason to believe that someone at the Florida meet would achieve two miles in under one minute. The "prettiest girl in Florida" was standing by to reward the driver who did so with a victory kiss. Race day dawned perfectly. With the weather cooperating, Marriott's toughest competition loomed in the person of Victor Demogeot, now piloting the 200-horsepower Darracq (Punnett 1997, 51). Marriott was the first to attempt to break a minute. He failed, posting sixty-three seconds in his run. Demogeot took his turn, bettered Marriott's time but fell short of the minute mark (61 3/5 seconds). In his next trial,

Marriott passed the minute barrier in a time of 59 3/5 seconds. But he hardly had time to gloat, as Demogeot bested him in the next run. Marriott appealed to the judges for one more trial in accordance with the regulations set for this race. But by the time Marriott got ready for his final try, the timing devices at the start line had been taken down. (Punnett 1997, 52). Race officials declared that Marriott had taken more than the allotted fifteen minutes to ready his automobile for a third attempt (*Automobile Magazine*, 1906, 278). The Frenchman's record stood: 58 4/5 seconds. Stanley was denied a sweep of the speed records, and steam aficionados cried foul.

It is hard not to see an element of prejudice in the decision of the officials to dismantle the timing apparatus so hastily. While it is true that Morgan himself, through the agency of his friend J.F. Hathaway, had issued a personal invitation for F.E. to compete at Daytona, it may also be true that as the racing events unfolded, Stanley's presence proved too much of a good thing. Morgan had hoped that a Stanley steam entry would heighten the competition and focus international interest on this American event, but he could not have wished this at the expense of the big name European contestants he had so painstakingly coaxed across the Atlantic. After all, by 1906 the internal combustion engine dominated automotive technology and consumer sales. Steam cars had slipped from favor with the motoring public, and few in automotive circles wanted to see steam cars triumphant. It is plausible, therefore, to conclude that the Ormond-Daytona organizers were not willing to risk the chance that the gasoline competitors would be humiliated again by Stanley's steam-powered "canoe." In addition, some of the wealthy spectators had bet a lot of money on the Darracq, especially after Stanley challenged its eligibility. The organizers may have feared the uproar that would have accompanied a third heat and a Stanley victory.

After the races were over, gasoline car proponents grumbled in print about the Stanleys' presence at Daytona. In an article for *Automobile Magazine* entitled "Carnival on Florida Beach," reporter A.G. Batchelder complained about the Stanleys' poor sportsmanship in the two-mile race: "…(T)he inventor of the freaky 'Teakettle' tried to put his gasoline opponent out of the game by questioning its eligibility on the grounds that it lacked a differential, an obsolete racing condition abroad, where the car was constructed" (1906, 271). *The Horseless Age* reflected a similar mood of displeasure with Stanley's domination of events and complained in their January 31 edition: "The Ormond meet …was even a greater disappointment to those who went to Florida to witness keen competition than the meet last year. Records were broken, but interest was entirely lacking." For its part, the *Scientific American* reported impartially, calling the Stanley showing at Orlando a "complete triumph of a steam racer built by the originators of the steam auto in America, the Stanley Brothers of Newton, Massachusetts" (1906, 115).

If the ruling powers in the automotive world were indifferent to the Stanley achievement, Newton nonetheless hailed the Stanley entourage, as Gustie put it, with a "hero's welcome." She celebrated her triumphant return home with an afternoon at the Boston Symphony and a new spring shirtwaist from her dressmaker, Madame Elise. Locally, Frank's racing success created a flurry of interest in his steam automobiles. People clustered four and five deep around his exhibit at the Boston auto show in March of 1906. Gustie recorded in her diary that "everyone was eager to hear Frank speak." Perhaps he promised them that the Stanley racer would return next year to Ormond.

Augusta Stanley watched Fred Marriott's disastrous run from the clubhouse of the Florida East Coast Automobile Association. Photograph by Edward G. Harris.

Stanley Museum Archives.

CHAPTER 13

Like a Meteor through the Surf

F.E. DID INDEED RETURN to the races at Ormond-Daytona in 1907. This time his stable of racing cars included the red canoe-bodied racer—essentially the same model Marriott had driven the year before but with an enhanced thirty-inch boiler and improved brakes; the two controversial Vanderbilt racers with elongated hoods for distance runs; and a Model K with a shorter coffin-nosed hood. The long-bodied racers were the same two designed but not delivered to compete at the Vanderbilt races on Long Island the previous summer.

The Daytona organizers clearly were not happy to see these Stanley entrants. Steam dominance in the speed events the previous year had indeed had a chilling effect on European participation. No Fiat, Darracq, or Napier cars came to Florida for the 1907 events. The only foreign entry that showed up to compete was a twenty-horsepower

Rolls Royce. For Morgan and the other organizers at Ormond this was an agonizing turn of events. They had struggled to win international recognition and now their efforts were going up literally in a cloud of steam. Because turnout for the competition was so meager, race officials reluctantly decided to trim the meet from six days to four.

F.E. Stanley, used to plying his own course in most things, appeared oblivious to the situation. He and Fred Marriott were bent on beating the marks Fred had set the year before whether any Europeans showed up to contend for them or not. The weather, however, was not boding well for the setting of records. An exceptional January dry spell had left the sand surface rutted and uneven in patches. Gustie wrote in her diary that the beach looked "unfit for racing." Picking up on the generally somber mood surrounding this year's events, she noted that there seemed "little interest in the races."

On January 22, the first day of competition, the Stanleys got off to an auspicious start as Frank Durbin in the twenty-horsepower Model F beat out a Rolls, and a Stoddard-Dayton to win the touring car event. Day two, however, proved a dismal one for the Stanley camp. During the ten-mile open race three Stanley steamers broke down—Fred in the red racer, Frank in one of the Vanderbilts, and Mr. L.F.N. Baldwin, in the other Vanderbilt. Gustie was "heartsick" as the three steamers were towed back to the shop for repairs. Luckily the mechanical problems were varied enough so that parts could be swapped among the cars. One Vanderbilt and Fred's racer, on which hung the hopes for smashing the mile record, were readied for the next day's competition.

The third day of events included the one-hundred-mile open and the first heats for the measured mile. The Stanleys hoped that their specially fitted Vanderbilt model (though still without a condenser) would be able to compete at the long distance. The year before there had been no Stanley car with enough range to attempt this race. Dreams of conquering yet another title, however, were quickly dashed. The Vanderbilt washed out of the race and a Mercedes went on to win the longest event of the meet.

146   THE STANLEYS OF NEWTON

Now the honor of the SMCC rested solely with Fred Marriott and a new record for the mile. But in his first run, he posted a disappointing 31 4/5 seconds—well off his record-setting pace from the previous year. Marriott had all night to replay this heat in his head, because he had to wait until the next and final day of competition for another chance to break his own record. That morning, the first of his remaining trials proved an even bigger letdown. Marriott clocked out a full second slower than he had managed the day before. In his next try, he improved his time to 29 3/5 seconds but still fell short of his old mark. Everything now depended upon the final run.

For this last try, Marriott was unwilling to hold back, despite the condition of the beach which had improved only slightly over the four days of competition. He and Frank had discussed driving close to the water because the beach surface was smoother there. But a misty fog rolling in off the sea forced Marriott to set a course well back from the water's edge. As he drove into position two miles below the start line, the crowd on the beach and in the reviewing stands grew excited. On February 1, the *Newton Graphic* reported that the gloomy boredom at Daytona dissipated as the spectators clapped and chanted in anticipation of a record-breaking run.

Marriott turned the red racer around when he reached the two-mile running mark. "Here he comes!" the crowd roared as the racer shot forward toward the start line, picking up speed as it neared the tape where "Senator" Morgan was holding the official time piece for the race. With open throttle, Marriott powered over the start line heading for what would be the fastest and shortest run of the day. Just seconds later, the large front wheels of the racer apparently struck a rut in the sand and for an instant lifted off the ground, giving spectators the impression that the Stanley was airborne. This was the moment when the racer's rear-weighted design became an awful liability. As the front end of the auto crashed to the ground, the car veered and sawed toward the water, then rolled over and over, breaking apart in the surf. The hood of the car was hurled ten feet into the air; the boiler rolled free of the wreck into the sea and the chassis lay on the beach in a heap of splintered wood. A cloud of steam rose so high that it was visible at the finish line three quarters of a mile away.

Glenn Curtiss, whose name is more famously associated with airplane history, was the closest and most informed eyewitness to the crash. Curtiss was waiting just behind the start line on a two-cylinder motorcycle in anticipation of the next event. He had both a clear view of the accident and a mechanical appreciation of what went wrong. However much time and fable have enlarged the account of events on that day, Curtiss's description remains the most compelling:

> As is commonly known, these steamers come to the start with a very high pressure of steam, saving it until the line is reached, then opening wide the throttle and fairly shooting over the line. This sudden spurt, together with the flat boarded surface of the bottom of the car and the fact that all the weight of the car is well back, taken in conjunction with a slight depression in the beach, formed, in my opinion, the true cause of the accident. The slight depression in the course gave the car (which was

Mrs. Stanley realizes her worst fears about speed and automobiles as she examines the wreckage of the Stanley racer.

Stanley Museum Archives.

provided with light springs) a toss-up. The sudden application of power assisted in raising the fore part of the car, which, as I mentioned is very light. The floor acted as an aeroplane—the car glided, with the rear wheels only on the beach. It then swerved sidewise, and when the front wheels again came in contact with the ground, it was headed toward the sea, the wheels of course went down and the car rolled over and over, breaking to fragments. The boiler kept on going, and rolled several hundred feet farther than the balance of the car, the escaping steam giving the appearance of a meteor rushing through the surf." (*Scientific American* 1907, 128).

Gustie, too, was a witness to the scene that day. Viewing the race from the clubhouse of the Florida East Coast Automobile Association, she recorded in horror how the car was "dashed to atoms" as it went into the water seemingly just "a cloud of steam." Shaken by Fred's "pale and bloody" face after the accident, she confided to her diary that "Frank and I are ready to collapse."

It is hard to believe that with the racecar shattered to bits Fred Marriott came through the ordeal generally in one piece and semi-conscious. Without a roll bar or a seatbelt, all he could do was crouch as low as possible in the Stanley as it tumbled along the beach. Like the errant boiler, Marriott came to rest in the surf. When a team of doctors and F.E. Stanley reached his side, Marriott was ashen except for a stream of blood running down his face. The *Newton Graphic* reported that his racing hood had been torn off and his scalp badly scraped. He complained over and over about his back but fortunately sustained no broken bones anywhere. He suffered a serious injury to his right eye that made doctors worry at first about his losing the sight in that eye. A shaken F.E. Stanley told the press, "All we can say is that Fred won't die." He then telegraphed Marriott's young wife to come down to Florida.

The Stanley entourage returned somberly to Newton on February 1, without Fred Marriott. There were no celebrations this season, only F.E.'s quiet resolve never to build another racing car. That decision apparently came after much changing of his mind and no small amount of entreaty from Augusta. Stanley told reporters on the scene in Florida that he would not rebuild his racer but assured the *Newton Circuit* that he intended to go back to Ormond next season with a new car. His determination may have waned, however, the longer Fred remained hospitalized. Marriott was finally released on March 1.

Automobile Magazine, reflecting the prevailing sentiment of the trade journals, had no ambivalent thoughts about the return of the Stanley racer. Reporter A.G. Batchelder did not restrain his contempt or rage in recounting the events at Daytona in 1907: "A steam freak ...spoiled the fifth meet on the famous ocean boulevard." Intoning that "...the day of sprint racing in automobile competition has passed," he dismissed the steam racer as having "scant utilitarian automobile value." He charged that its appropriation of center stage "discouraged the presence of cars built to go the distance, and caused the speed gathering to be a lamentable failure" (1907, 247-248).

As for the Ormond-Daytona organizers, they made sure that no steam racer would trouble their event in subsequent years. The next season, they decided to emphasize long distance races of one hundred miles or more. While not abolishing the shorter distance sprints, they set eligibility requirements that effectively eliminated steam or battery autos even from the mile run. Each competitor had to average at least sixty miles per hour over a distance of one hundred miles in order to enter any event (Punnett 1997, 72). Thus the reign of the steam car atop the racing world came rudely to a halt.

After the initial shock of Marriott's spectacular crack-up had passed and the ensuing controversy begun to subside, speculation grew as to how fast he was traveling when he crashed that foggy day. Over time the story of Marriott's last run grew into a tall tale, helped in no small measure by the gregarious Fred himself. The walls of his auto garage, located at 130 Galen Street just a short way from the SMCC factory, were hung with racing photographs and memorabilia from those early Ormond races. As an older man, he enjoyed recounting the events of that day when he almost succeeded in besting his own record as the

Before his ill-fated run Fred Marriott and F.E. Stanley pause for a photo with the racer as an unidentified guest tries out the cockpit.
Stanley Museum Archives.

fastest man on wheels. Marriott was among those who believed the Stanley racer had reached a speed in excess of 190 miles per hour before the calamity (a speed supposedly calculated by several professors from the Massachusetts Institute of Technology). He himself claimed to have seen the racer's speedometer register 197 miles per hour at the moment the front wheels left the beach (R. Stanley 1963, 128).

Raymond Stanley, however, bent considerable attention to keeping the record straight. He had been a thirteen-year-old boy when Marriott ran his disastrous race. Writing an account for the *Automobile Quarterly* in the summer of 1963, Raymond expressed doubt that the racer was equipped with a speedometer or that an accurate reading could have been taken even if such an instrument were on the car. He noted that his father had timed the race himself, "(snapping) off his watch the instant he saw the racer waver." The watch showed six seconds elapsed. Later in the day, F.E. measured the distance from the start line to the spot where the tracks indicated Marriott began to falter. The steam racer had covered a quarter of a mile in that six second run. This meant that with all other factors equal, Marriott would have completed the mile in twenty-four seconds at a speed of 150 miles per hour—an impressive feat even without hyperbole.

Would the course of automobile history have been different if Marriott had not crashed and the steamers had been allowed to continue to race? That is, of course, impossible to answer to a certainty, but likely it would not have been. While 1907 marked the best year in sales for the SMCC, the number of cars sold (seven hundred) represented a miniscule portion of the automobile market. Some motorists may have been swayed by the Stanley performances at Ormond-Daytona, but it seems unlikely that further steam racing triumphs would have significantly altered the mainstream of consumer trends. The average motorist cared less about how fast he could go than whether he got there without blowing up.

The fact that automobile boilers differed markedly from their industrial counterparts, which exploded with lethal regularity during the nineteenth and early twentieth centuries, seemed to make no impression on consumers. Industrial boilers had no automatic turn-off, so a leak in the return line meant that the water level continued to drop until the boiler dry-fired, causing an explosion (Technotes, 2001). Since steam cars came equipped with low-water gauges and shut-off valves, explosions were extremely rare, occurring only with older, poorly

The canoe body of the Stanley racer is eerily evident in the splintered wreckage on the beach.

Stanley Museum Archives.

maintained models. The one time the Stanley Brothers managed to blow up one of their own boilers was under experimental conditions at the auto factory. They put a boiler in a pit with a burner under it and got it to explode only after welding shut all of its heads and tubes.

If fear of explosion (however unjustified) were not enough of a deterrent for ordinary drivers, the level of skill needed to operate a steam vehicle may have been. Not only did a driver have to fire up the boiler on a steamer; but he also had to master a daunting array of gauges, pumps, and valves and keep them in working order. For most automobile consumers, the promise of great acceleration could not offset the drawbacks, real and imagined, of steam cars. Besides, gasoline cars were closing the speed gap by adding more cylinders to the internal combustion engine.

CHAPTER 14

From Racetrack to Roadway

IF THE STANLEY TWINS HAD BEEN DIFFERENT sorts of businessmen, they might have read in the snubbing of steam car racers a clear signal to find a new direction for their automobile company or to abandon production altogether. But the Stanleys, now in late middle age, showed no signs of following market trends. Just as they stuck to the business of perfecting dry plate after the photographic industry had moved on to roll film, so despite—or perhaps on account of—the ascendancy of the gasoline engine, they continued to manufacture their steam car.

By the middle of the first decade of the twentieth century the Stanleys survived, in modern parlance, as niche marketers—diametric opposites of Henry Ford, who placed a Tin Lizzy within the budget of every hard-working man. The Stanleys still eschewed mass advertising because they did not intend to appeal to a mass audience. They already knew the customers they wanted: no luxury hounds or folks looking for needless conveniences, but men like themselves who appreciated the basics of steam power. The brothers were, in short, mechanical snobs content to sell only to a small group of engineering cognoscenti who had the skill and patience to operate a steam vehicle. While Ford achieved high volume production, seven hundred autos was the maximum annual output ever reached by the SMCC. In its best years, the company turned only a modest profit, but the twins did not appear to mind. They had earned their fortune earlier in dry plate and in securities investments. The Stanleys did not get richer from building cars; they were in business for pleasure and pride.

In their dry plate days, the Stanleys managed to undersell their competitors by automating the coating process. By contrast, they did not apply their inventive genius to cost reduction in automobile manufacture. In fact, the longer the SMCC was in business, the more expensive its models became. For example, in 1903, the year the brothers redesigned the chain drive to direct drive, a two passenger 6.5-horsepower Stanley runabout—the most basic model that early automobile makers built— cost $650. In 1908, the Stanley EX ten-horsepower runabout cost two hundred dollars more. By 1912, a ten-horsepower Model 62 runabout sold for $1000. While none of these sticker increases seems surprising or excessive to the modern automobile consumer, in the highly competitive setting of a growing industry, they reveal the puzzling fact that the Stanleys were not using technology to cut expenses.

During the same time frame, Henry Ford was working hard to bring down the cost of his cars. He produced his first Model T autos in 1908. The base price was $850—the same amount the Stanleys charged that year for their runabout. By 1912, however, one year after introducing the automobile assembly line into his plant, Ford was able to offer a four-passenger touring car for $690, a roadster for $590, and a more luxurious Town Car for $900 (Henry Ford Museum). The assembly line made it possible to lower the price to the consumer by dramatically increasing the rate of production: between 1908 and 1927, the Ford Motor Company manufactured some 15,000,000 Model T cars. During its entire existence, 1903-1927, the SMCC built about 11,000 steamers.

Granting that the Stanleys had no more interest in doing business like Henry Ford than they had in matching the entrepreneurial output of George Eastman, it is curious to note that they did not compete within their own steam car niche. Of course, there were fewer and fewer steam manufacturers to compete with in the first decade of the 20th century. Still, from 1900 to 1910, the White Sewing Machine Company of

Cleveland, Ohio, built a formidable alternative to the Stanley. Although the brothers were ever at work fine-tuning their steamer, they never felt urgently obliged to answer White's technological advances.

From the first year of production, White offered fully automatic burner operation and automatic water flow to the boiler. One year earlier on Mount Washington, Freel and Flora had to jack up the rear wheels of their Locomobile so that the engine could pump water to the boiler. The next year, Locomobiles came equipped with a self-service hand pump to supply the boiler. By 1907, White featured so sophisticated a rate-proportioning control system that the only valve the driver had to operate was the fuel shutoff (Crank 1997, 14). As early as 1902, White steamers were equipped with condensers that recycled the exhaust steam to the water tank. Although White cars got three times the range of a Stanley on one gallon of water (Crank 1997, 15), the brothers did not put condensers on their cars until 1915.

How, then, in an already constricted steam market, did the Stanleys manage to stay in business? For those drivers who preferred steam, a Stanley was in some ways easier to operate because of its mechanical simplicity, and it cost a lot less than a White steamer. Indeed, the latter priced its line of cars decidedly at the luxurious end of steam transportation. In 1905, White advertised its six-passenger touring car for $3200, while that same year the closest comparable Stanley—the five-passenger, twenty-horsepower Model F touring car—sold for $1500. For another reason, White did not gain quite the same legendary reputation from its racing exploits as Stanley did. White steamers were renowned as superb hill climbers until factory-built specials were barred from such events (Crank 1997, 15). Throughout 1905, White's specially built "Whistling Billy" outmatched the big European cars that subsequently showed up to test their racing mettle in Florida. But a White racer came in second at Ormond-Daytona that year, and it was a Stanley that raced into the record books in 1906. It seems no accident that the SMCC's top production year (1907) coincided with Fred Marriott's heroics on the beach.

Almost from the very beginning of production, however, years before the glory days on a Florida beach, the Stanley name was enveloped in a shroud of myth and nostalgia. Thanks in no small part to early Locomobile advertising, the Stanley brothers, ornery, frugal New Englanders, were synonymous with the steam car in the American mind. The fact that there were two of these characters identically dressed and bearded served to enhance the myth. For a nation plunging headlong through the Gilded Age, the tales of homespun Yankee virtue that circulated about the Stanleys offered reassurance that some things might not change.

It was said the twins were so particular that they themselves whittled the wooden patterns used for casting certain parts and machinery. In fact they used lathes and power woodworking machines to make their molds. The no-nonsense Stanleys were supposed to have painted their cars only black, but in truth they favored "Russian Blue" and "Brewster Green" with a contrasting cream trim. The Stanley racer at Daytona was painted red as were many of the sportier CXes, EXes and ten-horsepower models.

One popular though groundless tale alleged that the Stanley cars ran so silently that steamboat whistles had to be affixed to the first models to warn unwary horses and pedestrians. While a Stanley operated relatively quietly by human standards, equine ears were more sensitive. The burner on a Stanley emitted a high-pitched whine that frightened horses and caused many to bolt (like the hapless produce hauler's nag in the fall of 1897). The most persistent myth about the Stanley Steamer claimed that the brothers offered $1000 to any man who dared to hold the throttle open for three minutes. The story went that a steamer attained its maximum speed in less than three minutes and could not continue to accelerate. So anyone who held open the throttle risked being dashed to splinters. In fact, a steamer's power was limited by the capacity of its boiler. A Stanley would eventually run out of steam with the throttle wide open (R. Stanley 1963, 122).

In the end, the quality of the Stanley product, more than myth or nostalgia, probably kept the brothers in business, however modest the commercial flow. The Stanleys built a reasonably priced steamer, pared down but mechanically sound and extremely reliable. Perhaps they did not keep pace with the automotive advances of the White steamer because, for a time, they did not have to. The most loyal Stanley customers apparently shared the brothers' imperative: Keep the steam engine simple.

By 1907, the basic Stanley model was uncomplicated but substantially better than the first popular Locomobile of 1899. F.E. had stirred the crowd at the Mechanics Fair Exhibition with a two-minute, eleven-second mile. Just eight years later Fred Marriott flew over the same distance on the beach in Florida in 28 1/5 seconds. Stanley cars now had a steering wheel in place of the Locomobile lever. The water pump to the boiler had been improved, and with a skillful driver at the helm, a 1907 Stanley could maintain good road speed, climb hills without a falter, and travel between fifty and one hundred miles before taking on water.

The brothers made only subtle changes to each year's models and never strayed too far from their basic design. Technical improvements generally resulted from their own motoring experiences and were road tested by the Stanleys themselves before marketing. Between F.E.'s trips to Maine in all kinds of weather and F.O.'s motoring adventures in the Colorado Rockies, they logged up an impressive amount of road time. Both brothers were partial to the sports car model in the Stanley line. The two-seater Model H, which sold for $1000, was called the "Gentleman's Speedy Roadster." Described in the company brochure as "graceful and rakish," its twenty-horsepower engine was designed for the likes of F.E. and F.O. who enjoyed running over the road (where the pavement permitted) at speeds from seventy to eighty miles per hour. Fred Marriott described the Model H as the "best all around car Stanley ever built, better than the Model K for hill-climbing and average high-speed work" (Marshall 1996, 22).

In 1905 the Stanleys brought out the first of their popular coffin-nosed models. The Model F touring car seated four or five passengers comfortably. Speed was not the primary objective in this Stanley touring car but power (twenty horses), range and stability. Although the engine was still geared directly to the rear axle, the coffin-nosed design housed the boiler under the hood. By 1909 the Stanleys built the powerful Model M, which cost $2000 but still sold for considerably less than a White steamer or gasoline Lozier. Advertisements for the Lozier mentioned discreetly that prices started "at $5000 and up." A custom-built Lozier sold for as much as $25,000.

In 1911, the Stanleys put a skin of aluminum over the wooden bodies of some of their models. The following year they brought out a thirty-horsepower, seven-passenger aluminum-clad touring car, the Model 87. As a result of this aluminum layer, the weight of the Stanley cars increased slowly but steadily each year. Interestingly, in 1911, Henry Ford stopped making his coaches from wood or aluminum and switched to steel bodies instead.

By 1913, all Stanley cars were finally equipped with kerosene burners. As early as 1904, other steam carmakers used kerosene to fuel their burners because gas was deemed too volatile. Since the automobile-buying public already feared the possibility of boiler explosions, most steam car manufacturers equipped their models with kerosene burners in an effort to allay popular concern. But because kerosene initially cost more than gas, the frugal Stanleys chose to use gas burners in their autos. Only after the price of gasoline went up in response to demand did the brothers switch to kerosene. Kerosene, however, is harder to vaporize. This meant the Stanleys had to lengthen and redesign their vaporizer to accommodate the heavier fuel (Hart 1992, 10).

Nephew Carlton Stanley, who had formerly managed the Montreal branch of the dry plate business for his uncles, proved himself to be a man of many skills by designing the new Stanley burner. In his final plan, fuel entered the combustion chamber of the burner through small drilled holes instead of the customary slots. This modification reduced the tendency of kerosene burners to whistle (C. Stanley 1950).

Fully eight years after the last steam car raced at Ormond-Daytona, the Stanleys built their first cars with condensers. Although the White Steam Car Company had been putting condensers on their models for years, true to stubborn form, the Stanleys had not felt compelled to alter and, in their estimation, complicate their design. Perhaps because they themselves did not mind doing so, the brothers thought it reasonable to expect a driver to stop for water every twenty to fifty miles on a long trip. In town, one could fill up at the village water trough but in the open countryside there were no designated water stops for automobiles. So the Stanley driver—with his steam-powered rubber siphon at the ready—had to find an obliging farmer (with friendly horses and no bull in the pasture) to beg the use of his watering trough.

The most amazing story of Stanley tolerance for the "water stop" belongs to F.O. In 1903 while seeking a cure for tuberculosis in Denver, Freel was unwilling to forego the pleasure of his runabout and had the sporty little car shipped to him by rail. In late June, he and Flora decided to move to summer quarters in nearby Estes Park. After seeing Flora off by conventional rail and stagecoach transport, F. O. planned to make the sixty-five mile, mountainous journey by auto. He was supposed to carry a passenger from Denver, but the man never showed up. F.O. seemed hardly fazed by this development and set off alone in his runabout. Traveling solo meant he had to shoulder all driving responsibilities, including manning the water hose. After an overnight stop near the town of Lyons, Freel made it to Estes Park safely, apparently no worse for the strain. It took twelve more years before the Stanleys put condensers on their cars.

If driving in the Rockies while suffering from tuberculosis did not move the brothers to reconsider their stand on condensers, what did? Given their stubborn history of refusing to keep up with the automotive Joneses, the answer is surprising. Most likely, competition from an upstart steam car builder pricked the Stanleys' pride enough to make them rethink the problem. Abner Doble, forty-two years younger than the twins and born in the age of the combustion engine, began building impressive steam-powered automobiles in Boston around 1910.

Abner and his brother John grew up in California tinkering with their father's White steamer. Like the Stanleys, they bucked market trends and immersed themselves in steam car mechanics. Abner spent a year studying engineering at MIT in Cambridge. Then in a move reminiscent of the young Frank Stanley, Doble left college to work on his own in a machine shop in Waltham, Massachusetts. During that time, according to Doble legend, Abner drove a steam car model with working condenser past the nearby automobile factory of the Stanley brothers. While the Doble version depicts the student taunting the old masters with his exhaust-free model, Raymond Stanley remembered things differently. He maintained that Doble applied for a job at the Watertown plant and that the brothers refused to hire him because they did not trust him. There's truth in both accounts.

While Doble went on to build a superb steam car that emitted not even a wisp of steam, the Stanleys, faced with dropping sales figures, got busy putting condensers on their cars. Frank was ready to road test these models by the summer of 1914. Gustie noted in her diary on July 9, 1914, that Frank had driven "in his new automobile" to the ferry for Squirrel Island (their summer home off the coast of Maine) and that "he didn't take water on the road for the first time in his life."

Following Frank's road tests, the 1915 Stanley line was redesigned. The twins replaced their signature coffin-nosed front with a radiator and condenser. Instead of escaping into the atmosphere, steam was recycled to the condenser where it cooled and was pumped again to the boiler. While the addition of a condenser doubled the range of the Stanley line, automotive historians generally agree that the brothers did not invent a condenser equal in performance to Doble's.

CHAPTER 15

All the Way up the Boulevard

DESPITE THEIR LONG-STANDING RESISTANCE to steam condensers, one imagines that in time the brothers themselves, now gentlemen in their late sixties, may have come to appreciate the extended range of their steamers. Like the rest of prosperous Americans in the early decades of the twentieth century, the Stanleys and their families had grown increasingly dependent on the automobile. They, in a manner of speaking, had been taken over by their own invention. They drove everywhere it was possible to do so at the drop of a motoring cap. From pleasure riding on a spring afternoon, to family trips to Kingfield, the Stanleys were on the automotive go.

Flora Stanley, who enjoyed handling a good trotting horse, apparently learned to drive a steam car as well. When interviewed by the Mt. Washington newspaper *Among the Clouds* about her drive to the summit with Freel in 1899, she commented, "A lady can very easily learn to steer...She needs to become somewhat familiar with the mechanism, but it is not at all complicated." Flora seems to have stopped short of assuming mechanical maintenance for the vehicle and advised prospective women drivers to "know enough about its workings to instruct the stable boys whom she would have to hire to take care of it."

Sister-in-law Augusta, by contrast, never learned to drive an automobile. Gustie did take a turn at the lever of Frank's earliest model, and with little Raymond watching in amazement, started to motor up the grand front steps of their home at 638 Centre Street (R. Stanley, August 24, 1955). Daughter Blanche had considerably more luck steering that first Stanley, but Gustie never tried again. As befit a matron of her social standing, she had a chauffeur and Stanley limousine at her disposal. One of her first drivers apparently drank on the job and as a result smashed the car into a tree with Gustie and Raymond on board. Luckily, only Gustie's pride suffered injury. As she noted in her diary on July 7, 1906, she dismissed the driver on the spot, preferring that day to walk home with Raymond securely in hand (Stanley Museum Archives).

After she finally engaged a reliable driver, the automobile became Augusta Stanley's preferred method of travel around Newton and Boston. In fact, she considered it a burden when she had to take the electric cars and complained that the train schedule to Boston was inconvenient. When she had trouble with her knee (which was quite frequently), she even shopped from her car. An old sledding injury, plus a bad auto wreck with Frank, dogged her in middle age with discomfort so severe she sometimes had to walk on crutches and even ride in a wheelchair. Naturally she was delighted when obliging clerks from Sawyer's department store brought shirt samples right to the curb, and salespersons from Jordan's enabled her to select the appropriate livery for Wallace, her favorite chauffeur, without leaving her seat. Gustie was an ardent Thursday afternoon Boston Symphony patron, and Wallace waited with the other chauffeurs to pick up his charge when the concert was over. She also had him on duty to take her to the numerous club meetings and bridge luncheons that filled her days. She admitted to her diary on May 29, 1907, that she felt "quite the thing (emphasis in the original) riding along in [her] limousine with Wallace in his best rig" (Stanley Museum Archives).

Every year from 1909 on, the Frank Stanleys summered at the elegant cottage they built on Squirrel Island in Boothbay Harbor, Maine. The latter represented a significant social step-up from the family's previous vacation retreat at the Preston Hotel in Swampscott, Massachusetts, but called for more laborious planning. Moving the Stanley household

Augusta Stanley's limousine: No wonder she felt herself "quite the thing" with handsome chauffeur Wallace behind the wheel.

Stanley Museum Archives.

to an island for the summer entailed a lengthy drive, caravan style, to Boothbay, then a ferry ride to the cottage. Gustie described one typical auto trek in June 1911 in which Frank's new four-seat "torpedo body" suffered repeated pump problems, while seventeen-year-old Raymond's red two-seater sustained two punctured tires. Flats and pump problems to the contrary, the Stanleys still preferred the freedom of automobile travel to the rigid schedule of the railroad. When the family moved for the summer, only the household staff traveled by train. The flexibility that the auto afforded seemed to make up for a host of inconveniences. The Stanleys, like the rest of motoring Americans, were content to travel with a tool kit, spare tire, and picnic lunch. A sense of humor didn't hurt either.

Many of the disabling woes visited upon drivers in the early 1900s stemmed from the atrocious condition of the nation's roadways. Most US road surfaces were still sand, mud, occasionally gravel or, at best, deteriorating macadam. Indeed, bad road surfaces made worse by rain produced the greatest travails for the early American motorist. The situation had gotten this bad during the nineteenth century because federal and state money earmarked for internal improvements went largely toward digging canals and laying railroads. Not until the 1890s did an effective national lobby for better highways take hold in this country. Led by bicycling enthusiasts (the League of American Wheelmen), civil engineers, carriage and bicycle makers, and eventually automobile manufacturers, the "good roads movement" resulted in dramatic changes in governmental policy. Even the railroad magnates briefly joined the push for better roads in order to improve access to their depots (Mason 1957, 2).

For a large part of the nation's existence, Americans had seen little reason to invest in good roads. The thirteen original colonies found it easier to trade with each other by water routes than to travel overland. After the Revolution, some enterprising souls began building private turnpikes with a little aid in money and land from their state legislatures. But the high cost of road upkeep, coupled with meager dividends for investors, turned most of these pikes into public wards (Mason 1957, 9-10). Townships took over the management and maintenance of roads with little distinction between thoroughfares and back routes. If a road traversed more than one township, each arm of local authority only took care of the stretch that passed through its jurisdiction. With little or no incentive, the science of road building in America remained primitive. Unskilled road crews, frequently consisting of farmers working off their road taxes, built and repaired the nation's highways and byways (Mason 1957, 30). In fact, the practice of exchanging labor for taxes proved so popular with farmers that it contributed to their stubborn stance against the good roads movement. Not until the federal government instituted rural free mail delivery in 1896 did farmers begin to convert to the cause of better roadways (Mason 1957, 202).

In the second half of the nineteenth century, the invention of the bicycle—in particular those refinements known as the safety bicycle and the pneumatic tire—touched off a national passion for mobility. The League of American Wheelmen (L.A.W.), organized in 1880, soon had its members on cycling tours all over the country. Appalled and bewildered by the state of the nation's roadways, these new travelers pressed state legislatures for better-marked roads and better road surfaces. The wheelmen came to realize, however, that before politicians changed their attitudes, ordinary citizens would have to change theirs. So the L.A.W. set itself the task of educating the public about the need for good roads. Central to that campaign was its weekly *Good Roads Bulletin* edited by the Stanleys' old friend, Sterling Elliott.

How extensively the Stanleys themselves participated in the lobbying efforts of the good roads movement is hard to assess. F.O. Stanley appears to have been more involved than his twin. The former belonged to the Newton Bicycle Club and while recuperating in Estes Park, joined the Colorado Good Roads Movement. F.O.'s longtime friendship with Sterling Elliott, coupled with his own compelling interest in the quality of mountain roads, doubtless led him to active involvement. At the state convention in 1906, Freel was chosen to serve as the first vice-president of the Colorado organization (Pickering 2000, 100). Frank Stanley's official participation included an appointment in 1900 to the National Highway Commission. His letters from abroad reveal how much he envied French auto makers their excellent roads, and there is evidence that on several occasions he delivered a few words—humorous and straight—urging the citizens of Maine to improve the quality of their country motoring.

By the mid-1890s, sentiment for road policy reform had gained national sway, and a number of states (Massachusetts was the second) established highway departments. This meant wider jurisdictions and deeper sources of revenue for building and maintaining roads. With the Massachusetts Institute of Technology leading the way, road building became a recognized engineering science. As a result, skilled crews began replacing statute labor (Mason 1957, 223).

Even the federal government got involved in highway improvement. In 1893, the Office of Road Inquiry was established as part of the

Department of Agriculture. Chief among its duties was building experimental stretches of roadway to test both engineering techniques and the properties of surface materials like asphalt and Portland cement (Mason 1957, 151). After the turn of the century, the emergence of automobile traffic posed a new challenge to the road scientists. They had to find surfaces that could withstand fast-moving vehicles. Automobiles churned up the top layer of macadam, thereby spewing crushed rock to the side of the road in a cloud of dust. By 1905, with automobile traffic an irreversible fact of American life, the Office of Inquiry grew into the Bureau of Public Roads.

In all events, improvement in American roads came slowly and did not keep pace with the speed of development in the automobile industry. Motor cars took to the nation's roadways before the roads could handle them, with the results that motorists suffered many mishaps on their travels. The Stanley auto trips to Maine stand as adequate example.

Gustie recorded in her diary one rain-soaked adventure in the spring of 1906. While driving from Auburn to Livermore Falls en route to Kingfield, it began to pour. The steamer was soon hopelessly mired in ankle-deep mud. Frank begged some rope from a farmer to bind around the wheels for traction. This arrangement worked until the rope wore through and they had to stop at another farmhouse for more rope. At Livermore Falls, Frank put chains on the car for the leg to Farmington. Gustie noted that they arrived at the family homestead "covered with mud."

On a spring trip to Boothbay Harbor in 1911, Frank and Gustie enjoyed a fine ride all the way to Kennebunk. Frank then decided to abandon the regular route for a drive through the woods, where they promptly got stuck in the mud. Gustie recorded the day's events in her diary on May 4, 1911: "Frank had to pry up the wheels—and I sat on the lever while he put old fence stuff under the wheels." Gustie's heft substituted nicely for a jack, and the pair made it to Portland "in time for dinner." It is hard to imagine the staid and fastidious Mrs. Stanley willing to endure such indignities except in the name of automotive adventuring.

So great was the appeal of the automobile, even in its grimiest early stages, that by 1910 the nation's biggest cities were already filling up with cars. Nothing less than a transportation revolution had occurred in the brief time since Frank Stanley built his first car. In 1897, there was no motor traffic in Boston or anywhere in America. There were only, in Samuel Eliot Morison's words, "a few experimental machines which were always breaking down" (Morison 1962, 28). When the Stanley Brothers began their automotive tinkering, livery stables of every description still covered the backside of Beacon Hill. Commonwealth Avenue, the long boulevard running from Boston through Brookline to Newton belonged to carriages and trolleys. Planners of that great thoroughfare could not have imagined that scarcely a decade later Commonwealth Avenue would teem with motor cars. Augusta Stanley herself bore witness to the transition one November evening in 1910 while she and Frank drove home from Boston. Looking out of the rear window of the Stanley, she observed that "the automobilists were just one continuous line all the way up the boulevard."

Newton Fire Chief Randlett in a 1903 Stanley.

Stanley Museum Archives.

CHAPTER 16

Beautiful Newton

THE AUTOMOBILE WAS BY NO MEANS the only transforming technology of the late nineteenth and early twentieth centuries. In one great rush, the telephone, the electric light bulb, the motion picture, the phonograph, all worked dramatic changes on the way people lived—particularly the wealthier classes in cities and suburbs. The extension of electrical power to private homes brought with it an array of appliances big and small that altered the rhythm of daily life. The well-appointed kitchen soon required electric refrigeration and a toaster, the modern laundry a washing machine and an electric iron. In 1909, Augusta Stanley hired an agent from Shreve, Crump, and Low in Boston to "go through the electric lighting" in her home—presumably to ensure that all fixtures were up to the latest standards. Gustie considered the purchase of her first electric vacuum cleaner (1910) and washing machine (1916) important enough to note in her diary. In the summer of 1912, she sounded sympathetically contemporary as she anticipated trading her busy life in Newton for the slower pace of Squirrel Island. On June 23, 1912, she wrote in her diary: "It seems good to get here once more, away from telephones, automobiles, and the thousand and one complications that go to make up our present highly complex manner of living."

Mrs. Stanley was not mistaken about the tempo of Newton life. After 1900, the city retained few markers of its simpler past. There were no more farms in either Newton Centre or Newton Corner, where the

With the advent of the automobile age the livery stable gave way to public garages.   *Newton History Museum, Newton, Mass.*

Newton Fire Chief Randlett in a 1909 Stanley. The use of steam pumpers (center) led many fire chiefs to continue to rely on steam cars. Here Newton Fire Chief W.B. Randlett is a passenger in a 1909 E2 Stanley (right) in front of the fire department's Company 3 Station.

Photo courtesy of Allen & Janette Blazick.

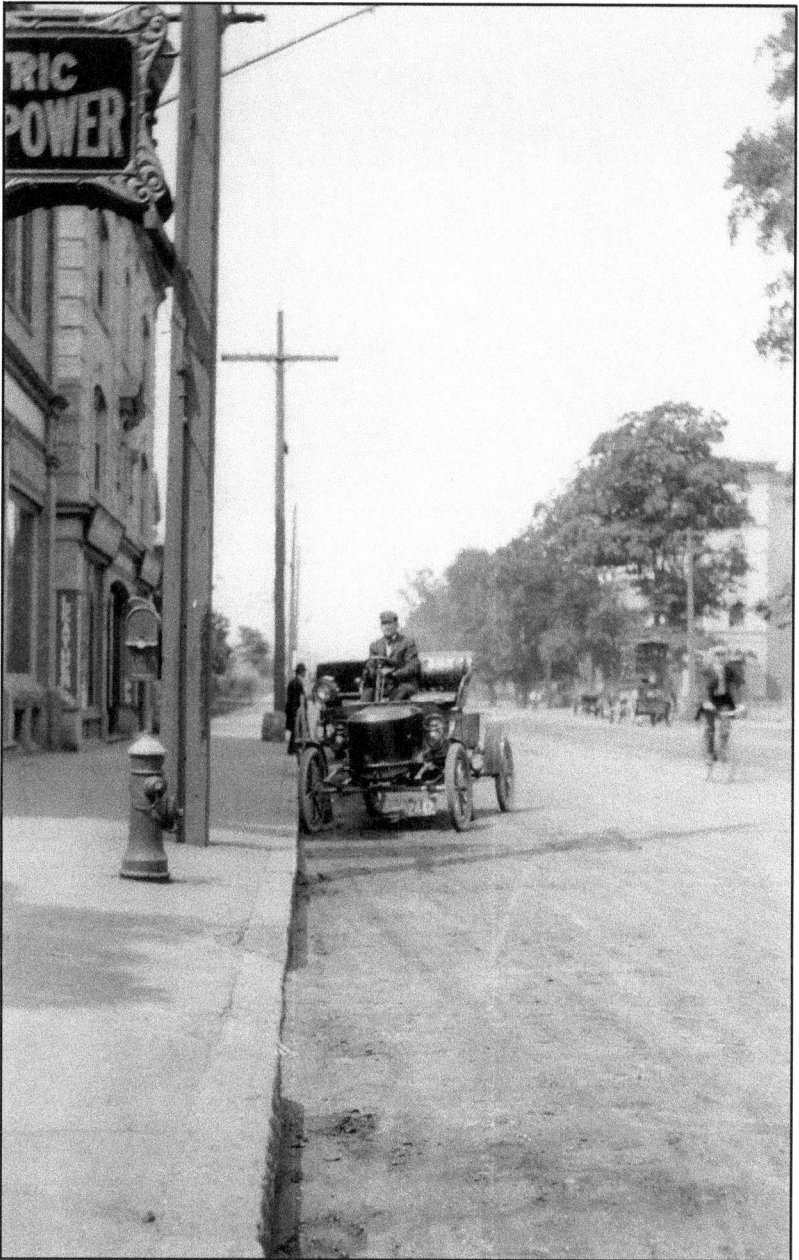

An early Model E Stanley runabout parked on Washington Street in
Newton Corner. The Model E, built from 1905-1907, was Stanley's popular
low-cost model, with more than 750 built in three years.

Newton History Museum, Newton, Mass.

Gateway to Newton: Newton Corner in the age of the electric trolley and the automobile.

Newton History Museum, Newton, Mass.

Stanley brothers lived. Even farmers' markets set up in these two villages failed to win over housewives from the convenience of their green grocers (Rowe 1930, 392). Retail stores lined the square in Newton Corner and a stone bank building anchored the intersection of Washington and Centre Streets, across from the Newton Free Library. Down Galen Street into Watertown, the houses of working men were densely clustered, and commercial buildings—including the Stanley Motor Carriage factory— stood along the Charles River. Crossing Newton Square on foot required less courage now since the tracks of the Boston and Albany Railroad had been depressed. During the 1890s, pressure from neighborhood improvement associations had finally led to a 2.25 million dollar project (largely funded by the railroad) to lower the railroad tracks throughout the city and thereby eliminate dangerous grade crossings.

Even leisure time had gotten more organized in Newton. The Charles River had always been a popular spot for recreation, but now a park, laid out along its banks by the company that ran a trolley line down the center of Commonwealth Avenue, offered day trippers and residents alike a new destination. For 25 cents roundtrip, the shop girls and store clerks of Boston could ride to Nurembega to stroll the nature paths, visit the zoo, rent a canoe, hear a band concert, or eat in the park's Pavilion Restaurant. Commonwealth Avenue itself became the smart new address in Newton and building boomed along its wide expanses.

As the twentieth century unfolded, Newton was growing increasingly electrified and motorized. By 1902, the proliferation of telephones in city homes necessitated the building of a large new telephone exchange in West Newton. Electrical street lighting replaced gas lamps. Horse-drawn trolleys had already given way to electric ones, and now automobiles began to supplant public conveyances. Since that September day in 1897 when the Stanleys' experimental motor car supposedly put a local workhorse to flight, the automobile had become a commonplace sight in Newton. By 1916, the city assessed 2000 cars within its limits (Rowe 1930, 388).

The growing automobile culture occasioned certain architectural innovations around the city. Gas pumping stations popped up and the family garage, often a converted horse barn, made its first appearance. Many automobilists without a stable to remodel housed their cars at places like Fred Marriott's garage. Still others squeezed a garage and driveway next to their house. The addition of a garage to property not intended for such an appendage prompted some clever solutions. When F.O. Stanley, for example, built a garage next to his first Newton home on Hunnewell Avenue, he put up a small structure with a wooden turntable inside. Turntables were long familiar from railroad switching yards, and in Kingfield, the Stanleys' older brother Isaac had installed one in the family barn. F.O. parked his steamer on the turntable in his garage, and when he was ready to go out manually pushed the car around so that it was facing front. When he and Flora moved to a larger home at 337 Waverly Avenue, they presumably had ample space to accommodate their automobiles but nonetheless stuck to the turntable design. Maybe F.O. simply did not like backing up.

By 1900 the Stanleys had lived in Newton for a decade. Like the Garden City itself they had shed most outward vestiges of their rural

This undated photograph shows the steps leading into the music room at 638 Centre Street. *Stanley Museum Archives.*

Mrs. Stanley frequently redecorated at 638 Centre Street. This is an undated view of the parlor. *Stanley Museum Archives.*

beginnings. The financial success of the Stanley dry plate business in the 1890s had boosted the twins steadily from the status of comfortably fixed to the near precincts of opulence. The house that Frank Stanley built in 1896 reflected that rise in means. It stood at 638 Centre Street on one of nine contiguous lots the brothers had purchased together from the breakup of the old Hyde estate. The property lay between Gardner Road and Hyde Avenue. Eventually Frank and Freel made an even swap of real estate so that Frank owned all nine lots. According to Augusta's diary entry on June 5, 1907, Frank took over his brother's share of that parcel (worth $10,000) by relinquishing half ownership in another of their joint real estate holdings—the Hunnewell Club in Newton Corner.

Frank designed the colonial revival house himself with the help of a Boston architect named Emerson. Although the house is no longer standing, photographs reveal the elegance and symmetry of its design: Ionic columns stood at the corners; a rounded portico with balustrade

The living room at 638 Centre Street.   *Stanley Museum Archives.*

The dining room at the F.E. Stanley residence.   *Stanley Museum Archives.*

marked the front entrance; a large Palladian window motif was situated over the entrance. There were dormer windows and a roofline capped with a widow's walk. In 1916, Gustie herself oversaw the addition of two sunrooms. The house was originally built with a white clapboard facade, but Frank had it refaced in 1912. After constructing an automobile factory and a home for each of his daughters in reinforced concrete, F.E. figured out a technique for "slapping" a coat of concrete on his own home.

Inside, the domestic appurtenances included a billiards room, bowling alley, and a music room of "performance quality" (B. Hallett 1954). As a young girl in Lewiston, Blanche Stanley had thought the Dingley's music room with its reproduction statues and grand piano the most wonderful space she had ever seen. But its amenities clearly paled along side her parents' "theatre-like room." To reach this chamber, one climbed thirteen stairs to the second floor landing. There the stairway curved to the left and the right. Two broad steps in the center of the landing led to the music room, which was slightly lower than the second floor. The ceiling of the music room, painted pale blue with fleecy clouds and cherub musicians, was open up to the third floor.

Over the years, this room was the site of many chamber music concerts and recitals, called musicales in the day and attended by as many as a hundred guests at a time. The performers were usually prominent local soloists, but on one occasion a young Italian violinist named Geruidi performed a solo on an instrument Frank himself had made. Even their daughter, Blanche Stanley Hallett, who had studied voice for a number of years, and her talented violinist son, Stanley, were asked to perform. Gustie was particularly fond of the agent for the SMCC in Britain, one Mr. Ernest Galloway, because he had a fine singing voice and liked to join the family around the Steinway. Galloway's business trips to Newton often occasioned a musicale in his honor.

The garage at 638 Centre was perhaps as noteworthy as the music room. It was a three-story structure that housed the cars of Frank and his

The concrete houses built by F.E. Stanley for his daughters on Hyde Avenue: Blanche's house on the left and Emily's house to the right.

Stanley Museum Archives.

The front of Emily Stanley Warren's cement house at 35 Hyde Avenue.
Stanley Museum Archives.

sons-in-law on the first two floors and had a ballroom on the third. The Stanley grandchildren took dancing lessons there; Blanche and Emmie gave chic parties; and two granddaughters, Augusta Hallett and Emily Warren, had their coming-out parties together in the family ballroom in 1927. At holiday time the garage was often the scene of a family reunion.

Stanley gatherings included not only a contingent from Kingfield but also a cluster of relatives who had been drawn to Newton—some temporarily, some permanently—by the twins' manufacturing success. These included their youngest brother Bayard; Isaac Stanley's sons, Newton and Carlton; plus a handful of cousins. Augusta Walker's unmarried sister Emma, who worked first in the dry plate business and later at the SMCC, was also among their number. The most extraordinary of the lot was the twins' only sister, Chansonetta Emmons. Widowed in 1898, she moved from the Dorchester section of Boston to Newton with her eight-year-old daughter Dorothy.

Chansonetta was as talented artistically as her brothers were mechanically, yet her achievements remained largely private during her lifetime. As a young woman she left Kingfield with parental blessing to study painting at Western State Normal School (now the University of

This rural interior is typical of Chansonetta Stanley Emmons' domestic dramas. Her own daughter Dorothy is playing with corncobs while Tristram Norton shells corn. Photograph by Chansonetta Stanley Emmons.
Stanley Museum Archives.

Maine at Farmington) and for a time taught art in Boston. During the 1880s, F.E.'s portrait art and photographic endeavors in Lewiston caught her attention. In 1897, a year before her husband's death, Chansonetta undertook her first serious venture into photography during a trip to Asheville, North Carolina. Once widowed, she divided her time between caring for her daughter and mastering the photographic art.

Although she shot some fine architectural pictures and landscapes, the best of her photographs capture ordinary people at home and at work. Her New England images, taken around and inside simple farmhouses, depict men and women engaged in the kind of repetitive tasks once necessary to sustain life but in the early twentieth century rapidly slipping into desuetude. Ironically, Chansonetta attempted to preserve in photographs a pastoral world made increasingly obsolete by her brothers and other innovators of the technological revolution. But her pictures are neither sentimental nor frozen in time. Through the arrangement of detail and the implied relationships among or between the sitters she presents domestic drama not mere tableau. Her scenes take on emotional intensity from the contrasting play of light and shadow on ordinary objects. She achieved the latter effect by "dodging" or "burning in" certain elements as she developed her own glass plates (R. Stanley 1990, 12).

During the 1920s, Chansonetta traveled beyond the rural familiarity of New England with a grand photographic tour of Europe and a return visit to the South. While touring Charleston, South Carolina, she chronicled in pictures the working life of African-American women, several years before the US government commissioned such documentation from the Federal Arts Project (Gardner-Huggett 1996, 8). Chansonetta's close ties to her brother Freel and his wife Flora took her many times to their Rocky Mountain retreat in Estes Park. Some of her Western landscapes presage the luminescent monumentality of Ansel Adams' work.

According to Raymond Stanley, Chansonetta was asked to display some pictures at the Boston Museum of Fine Arts at least once during

Chansonetta's view of the Rocky Mountains is both mystical and monumental. Photograph by Chansonetta Stanley Emmons.

Stanley Museum Archives.

her life, but she did not make a living from selling her pictures. Photography—apart from portraiture—was not yet coveted in American homes nor indeed fully recognized as an art form. As a result she had to be content with small commissions from family and friends (Pearson 1992, 6).

After her husband died in 1898, Chansonetta became financially strapped and fully dependent on her brothers. Since Frank paid a considerable amount for the upkeep of his daughters' families, Freel assumed payment of the rent on Chansonetta's residence at 21 Bennington Street in Newton Corner. Perhaps because he and Flora were childless themselves, Freel spent more time with his sister than did Frank. But it did not help that relations between Augusta Stanley and Chansonetta tended toward contentiousness. Gustie once deemed her artistically gifted sister-in-law the "most uncomfortable person I ever ran up against," and evidence suggests that Chansonetta returned the sentiment. On several awkward occasions, Frank was called upon to mediate, however unwillingly, between his wife and sister. But at holiday time, as in most families, squabbles gave way to festive celebrations.

On Christmas Day, 1909, thirty Stanleys gathered at Gustie's home for turkey, goose, roasted oysters, Waldorf salad, flaming plum pudding, pie, ice cream and cake, figs and candy. After the feast they repaired to the ballroom for dancing, fiddling, and blind man's bluff. That party apparently proved so successful that two Christmases later, Gustie staged the entire affair in the garage ballroom. With pine bows festooning the walls, a fire in the fireplace, and tables laden with cakes, twisted molasses doughnuts made by Gustie herself, popcorn balls and ice cream, the Stanley clan gathered to dance and play cards.

The Centre Street house was the site of the one wedding Augusta got to preside over. Emma Stanley had been married in quiet haste in 1896. According to her sister Blanche, F.E. wanted the union annulled, but, because Emma was pregnant, the family did not interfere with the marriage (B. Hallett 1954). It must have been a tremendous blow to her parents, but Emmie clearly was not banished from grace. In fact her

Isaac Stanley's wife Minerva churning butter is the subject of this pastoral photograph. Photograph by Chansonetta Stanley Emmons.

Stanley Museum Archives.

husband Prescott Warren was eventually brought into the management of the SMCC, and, as a married woman, Emmie was her mother's frequent and cherished companion on shopping tours and at the matinees.

Having cruelly missed the pleasure of staging one wedding, Gustie was not to be denied when her older daughter finally decided to marry. Blanche's wedding to Edward M. Hallett on October 14, 1903, was in the bride's estimation, "too elaborate an affair," but one that "pleased [her] mother." Six bridesmaids in Paris-designed gowns attended the bride; Gustie had a hairdresser on call to do the women's hair. Despite a demurral from Frank himself who was never comfortable with organized religion, the evening ceremony took place at the Channing Unitarian Church in Newton. After the service, the wedding party—even Uncle Freel and Aunt Flora had come back from their first autumn in Denver—and a large contingent of guests repaired to 638 Centre Street. For the comfort of all, an awning had been stretched from the curb to the front entrance. Inside, Gustie had spared no expense for food or decorations. In the drawing room under a huge arch festooned with pink roses, the bridal couple received their guests.

As a wedding present, Frank Stanley bought a first home for the couple at 694 Centre Street. But he had bigger plans in mind. He intended to build houses for both of his daughters on two of the lots purchased from the Hyde estate. The following spring, Frank hired the Boston architectural firm of Chapman and Frazer to draw up plans for two Classical Revival, English Arts and Crafts style houses built of reinforced concrete.

Residential building with concrete was something of a novelty in 1908, and Frank Stanley probably knew as much about it as anyone. According to Raymond Stanley, the architects conferred closely with Frank during the design and construction of the two houses. Five years earlier, F.E. had been among the first to use the material in his Watertown automobile factory. Satisfaction with the SMCC building apparently convinced him to employ concrete in domestic design. The walls of the houses were made of solid concrete panels divided by vertical wood posts.

Construction went forward during the summer and fall of 1908 for the Edward Halletts at 35 Hyde Avenue and the Prescott Warrens next door at number forty-five. F.E. assumed the project's full cost, which, as Gustie confided to her diary, tended to be higher than anticipated because the "girls want(ed) better houses than I supposed." When the Stanley daughters were fully settled in their new homes, however, Gustie wrote that she was pleased to have lived to see them "so well fixed in life." But she did continue to fret over the extravagance of their household expenditures.

Over the years, the F.E. Stanley home was the scene of many formal parties that Gustie detailed in her diary, from guest lists to menus. On several occasions she noted the wine selections as well. Her entry for January 5, 1907, mentions that she served claret, champagne, and creme de menthe at a dinner party. Apparently, a bit of Parisian influence still lingered in the F.E. Stanley household, though abstinence still reigned at the F.O. Stanleys. In her various capacities as a club woman, Gustie hosted innumerable ladies' bridge luncheons, but soirées were her favorites. Amid protestations that she was nearly exhausted by the effort, she generally confided to her diary great social success after an evening of entertaining.

Oftentimes, she gave the most elaborate party of the year on New Year's Day, the Stanleys' wedding anniversary. On January 1, 1913, the occasion of their forty-third anniversary, Gustie and Frank held a formal reception at their home between four and seven o'clock, hosted a dinner to follow for twenty-four guests, and capped the evening with auction bridge. "I can't begin to describe the party," she wrote in her diary, "there were roses everywhere and many gifts from friends." A rather wistful note marking the occasion arrived from their old compatriot Frank Dingley, who apparently had not been invited to the party:

> Irrespective of whether we are invited to your feast tonight, we shall be there in spirit...Harking back more years than I care to count, we observe an adventurous youngster...with his enterprising wife, camping in Lewiston...Through co-operation unusual and intelligence equally marked, this young couple gradually commanded what the world called success...achieving more than livelihood...breaking with the chimney corner in Auburn and moving on to the Hub...now in the heyday of industrial and social victory, incredible in the remote country of their origin, is the romance of truthful history (Stanley Museum Archives).

It is difficult to say why Augusta did not invite the Dingleys. While his note hints of friends perhaps outgrown, it may not be fair to conclude that Augusta snubbed her oldest chums for the sake of social standing. It is fair to say, however, that she paid considerably more attention to rank and status than did her sister-in-law Flora, who was more apt to note in her brief diary jottings that she had spent the day putting up preserves than hobnobbing with swells.

Gustie, by contrast, reported lengthy and critical accounts of the parties she attended and the people she met. After attending a bridge luncheon at the home of a new neighbor, she observed that it was "a pretty party," but the "real society element of Newton was conspicuous by their absence." She concluded in her diary entry for February 20, 1912, that newcomers often "make the acquaintance of the wrong people first." Sometimes Gustie's social ambition struck an inadvertently comical note. A bridge luncheon at the Somerset Hotel in Boston occasioned this diary notation on April 11, 1907: "It seemed rather good to meet with the fashionable ladies [emphasis in the original] once more...they are just as pleasant and nice to be with as those who wear dowdy gowns...."

It is easy to mock Gustie for her determined pursuit of the socially right thing, but her behavior was by no means unique in that day. To one degree or another, all the members of the newly minted technocracy were upwardly striving, no doubt because nothing in their previous experience had prepared them for the direction or the bounty of their adult lives. As a result of their uncertainty, the newly rich followed a detailed and rigid code of conduct based on imitation of upper class customs. Invitations to parties, for example, were always written, and it was unthinkable not to respond in kind with an acceptance or regrets. The matrons of Newton did not drop in on each other for impromptu visits, even if they were next door neighbors. A lady entertained visitors during her "at home" hours on a given day of the month and went to friends' houses when they were receiving or when she wished to leave her calling card. Once when Gustie was sick, she entertained guests from bed in her kimono so as not to miss her Monday audience.

Patronage of the arts ranked very high with the new middle class, therefore one had to have a regular seat at symphony. When the Boston Opera House opened in 1909, weekly attendance there became a must, too, though never on a cut-rate night. Gustie once turned in tickets when she realized that she had purchased them for a popular price performance because the bargain seats brought in "the wrong sort of audience." With Emmie and Raymond in tow, Gustie visited Boston's

**Raymond Stanley as a corporal
in the Signal Corps during World War I.**
Stanley Museum Archives.

The Stanley daughters, Blanche (left) and Emmie, in an 1892 photograph.
Photograph by George H. Hastings.
Courtesy Thelma Chenowith, Stanley Museum Archives.

new Museum of Fine Arts on Huntington Avenue and reviewed the trip
in her diary on February 27, 1910: "A very beautiful building," she aptly
wrote, but "a scarcity of pictures." A few years later, at a reception held
at the Guild of Boston Artists, Gustie met John Singer Sargent, while he
was in town to paint the murals at the Boston Public Library.

While the pace of Gustie's cultural consumption (Frank was not
always in tow) may have been dictated by fashion, her appreciation of
music was genuine and sophisticated. At Christmastime in 1910, Gustie
enthusiastically reviewed her first encounter with Dvorak's *New World
Symphony*. She claimed Sibelius as her favorite modern composer but
later grew enraptured with Wagner after a Boston Opera performance of
Die Meistersinger. On April 22, 1916, she wrote in her diary: "I think I

F.E. Stanley took this portrait of his sister Chansonetta in 1888. Photograph by F.E. Stanley.

Stanley Museum Archives.

never heard anything so beautiful as this Wagner music." Interestingly, she also found room for the less accessible compositions of Richard Strauss. She counted his *Also sprach Zararthustra* and the opera *Don Quixote* among her favorites.

Diligent weekly attendance at both the symphony and the opera rewarded her with audience to some of the great moments in Boston's musical history. On October 12, 1906, she attended the debut of acclaimed Boston Symphony Orchestra conductor Karl Munch. In April 1907, she heard Enrico Caruso perform in *Aida*. Although she called him "Padereski" in her diary, Gustie caught Polish piano virtuoso Jan Ignace Paderewski in one of his rare concert outings in Boston on November 15, 1907. Since Gustie frequently misspelled the names of foreign artists, it is likely that a March 5, 1915, diary reference to "the fine cellist Cassal" means that she attended a concert by the young Spanish master, Pablo Casals. She heard Fritz Kreisler, the Austrian violinist, in concert with the Boston Symphony in 1910. Eight years later, she and Frank both attended a performance (March 29) by eighteen-year-old violin prodigy Jascha Heifetz.

Attendance at the theater was socially mandatory for the circles in which the Stanleys moved. One took in the latest light comedy at the Hollis or the occasional performance by the likes of Ethel Barrymore and Isadora Duncan. Seeing the latter in dance recital at Symphony Hall in November 1909, Augusta wrote in her diary that "in some parts (it was) the most beautiful thing I ever saw." It is interesting to note, however, that Gustie generally did not maintain the same highbrow standards for theatergoing that she rigorously followed for musical performances. In fact, she and Frank seemed to prefer frothy theatrical fare to drama. According to Gustie's diary, Frank's taste traveled all the way to broad farcical comedy. Noting on December 5, 1914, that they had attended a performance of *The Crinoline Girl*, Gustie wrote that this was Frank's "kind of a play." Billed as a "farcical melodramatic comedy" with a ragtime score, *The Crinoline Girl* starred a female impersonator named Julian Eltinge in the title role (Merrick 2001, 6). The Stanleys hardly ever went to the moving pictures but they did see D.W. Griffith's *Birth of a Nation*. Gustie wrote in her diary entry for April 10, 1915, that Frank thought it the best thing he had ever seen in a theater. Objecting to the Southern sympathy evident in the picture, she was more reserved in her appraisal and noted the film had been deemed racist in New York City.

It was important for a woman of Augusta Stanley's social position to read copiously, and she did so with relish. Gustie took equal delight in the latest potboiler or a novel by her avowed favorite author, George Eliot. Literary lectures were in vogue and Gustie attended regularly. After one such talk, she resolved to buy a full set of Jane Austen's works. On the lighter side, Gustie subscribed to three magazines. In 1910, for a total of $5.40 per year, she received *McClure's*, *Harpers*, and *The American Magazine*, a popular illustrated monthly.

Stylish ladies like Augusta Stanley lunched at the English Tea Room and played cards at the Somerset or the Hotel Vendome in Boston. Men and women both played bridge, but the latter in particular cultivated their skill at cards the way earlier genteel generations perfected their needle work. Gustie, for example, took a class in auction bridge when it became the rage and confided embarrassment to her diary because Frank had not taken time to become a better bridge player.

This great surge in card playing corresponded to the sudden freedom of middle-class women from the daylong drudgery of keeping house. The wives of prospering technocrats found that their household duties now mainly consisted of hiring and overseeing a small domestic staff. Gustie herself employed two downstairs maids, one upstairs maid, a cook, and occasionally a laundress. If these middle class wives made forays into their own kitchens, as the Stanleys did, it was to prepare a special family recipe or exhibit some homemaking art such as baking or canning preserves. Relieved of the most onerous household chores, they happily settled into a round of socializing that took them beyond their own drawing rooms. The bridge luncheon, which afforded an excellent opportunity to see and be seen by the right people, quickly formed one pole of the suburban matrons' world.

Club membership, with its more didactic mission, formed the other. Well-to-do women in Newton and other select suburbs across the country organized as eagerly into clubs as they gathered at the bridge table. They applied themselves with the same energy and a great deal more earnestness to the task of mastering culture and solving society's ills. The impulse for self-improvement was hardly new on the American scene—debates, lectures, and library societies appeared in the late eighteenth century as a means to uplift community life. But the Gilded Age clubs had the collective effect of bringing large numbers of women together for the first time to study and discuss a range of topics from the fine arts to social welfare to female suffrage. (As to a woman's right to vote, neither Flora nor Gustie was a suffragette, though the former favored it and the latter had deep reservations).

The prototype for many of the clubs was provided by the Chatauqua movement of the 1870s, which purported to offer all comers, male and female, a college equivalency education at home. Newton had its own Chatauqua chapter in the village of the Highlands, but socially prominent women found the prospect of downing large doses of the arts and sciences among their select peers a more appealing prospect. By the late 1880s, elite clubs, with membership restricted by sex and limited by invitation, sprang up throughout the villages of Newton. A sampling of these clubs included the Newton Centre Reading Club, the Women's Educational Club of West Newton; the Monday Club for Women in Newton Highlands; and the prestigious Social Science Club (still extant!) in Newton Corner.

The format for these limited membership clubs was almost always the same: the group gathered regularly in private homes to hear members deliver papers on assigned topics. If Gustie's experiences provide an accurate gauge, the women took the preparation of their papers very seriously. According to her diary, she struggled through numerous drafts and many consultations with Frank before producing a talk on Rembrandt for the Social Science Club. She confided her insecurities to the pages of her diary, writing on October 21, 1907: "How I wish I could write as well as I can imagine. Oh! If I could have the education the modern girl thinks so lightly of, what a help it would have been to me in life." Gustie had an easier time when asked to relate events at the Vanderbilt Races and to give account, along with Flora, of her automobile adventures on Mount Washington. When Flora Stanley lectured and displayed her private collection of antique Chinese pottery to the Social Science membership, Gustie noted that Flora "had pleased her audience." Though in Gustie's opinion, dutifully recorded in her diary on March 10, 1909, the topic was "too big" to be covered in "so short a time."

A number of Newton organizations entertained civic as well as educational missions that drew the women from their homes into community service. The Social Science Club, for example, was dedicated to no less lofty a goal than the rational ordering of society. To that end, the membership sponsored technical-vocational training for the children of working class parents in Newton. The Nonantum Vacation School, which the women financed in partnership with city government, offered home economics for girls and shop for boys. However class-conscious this undertaking appears to modern eyes, the project was popular and grew to a point that warranted use of the Stearn's School building. Originally a summer school venture, the industrial school evolved over time into the Technical (vocational) High School in Newtonville.

Both Stanley wives belonged to numerous clubs, but the need to spend many months in Colorado for F.O.'s health limited Flora's activities in Newton. Still, she managed to find time for the Watertown Women's Club (a holdover from her days on Maple Street); the Katahdin Club (which she regaled with a paper on the first "Climb to the Clouds"); the Sarah Hull Chapter of the Daughters of the American Revolution; and La Ligue Litteraire, a reading club for men and women that drew its membership from metropolitan Boston. When she and Freel were in Denver, Flora apparently took up a similar round of activities with her Western sisters. (Pickering 2000, 53).

Not surprisingly, Gustie's Newton club life was more robust than Flora's. She, too, belonged to the Sarah Hull Chapter of the D.A.R. and the Social Science Club, serving as chairman at least once. She joined the Impromptu Club, which met for concerts. She was a member of the District Nurses Association and was president for two terms of that organization whose mission it was to provide home nursing visits to indigent patients. The group also sponsored lectures on medical and public health topics. Indeed, Gustie marked her reelection to the presidency with a guest lecture on the then popular topic of eugenics. Regrettably, she left no opinions in her diary on the genetic ordering of society.

Gustie's favorite club, however, was the Wednesday Morning Club because she felt it combined the proper mix of intellectual and socially prominent women from greater Boston. Gustie presented her Rembrandt paper to the Wednesday Morning ladies and happily recorded in her diary that her lecture had been a success. She brought her audience to tears with the pathetic tale of the painter's life. In May 1909, Gustie accepted the presidency of the club with a good deal of trepidation, admitting to her diary that she was "appalled at the idea of being president of such a club" whose members "are _so_ (emphasis in the original) cultured." After a few nervous outings at the helm, she began to enjoy her post but refused the nomination when asked to serve another year. In 1916, Gustie joined the Presidents' Club, a purely social group of past and current presidents of a federated club.

The proliferation of women's clubs during the early twentieth century, not just in Newton but across the country, led to state and national federations. The federated club women of Massachusetts sent delegates every year to a state convention, while the National Federation of Women's Clubs gathered biennially. Although they never traveled together to the same convention, Flora and Gustie each served as delegates at both the state and federal level. On the eve of one such national gathering in New York City, Gustie anticipated events. She wrote in her diary on May 22, 1916: "I predict a very grand Biennial with everything for the uplift of Humanity (sic) presented and ably discussed." Gustie could hardly have been disappointed when she arrived at the convention. The delegates heard lectures on immigration, conservation, home economics, and industrial working conditions. The Police Commissioner of New York City and the warden of Sing Sing prison addressed the assembly. Gustie got a chance to visit a night court session and witnessed twenty young women "brought in from the streets." Despite a full roster of events, Gustie and her friends managed to lunch at Gimbels and squeeze in some sightseeing, traveling by train to Edison's laboratories at Menlo Park.

In addition to giving women from all across the country a forum in which to examine the issues of the day, these national club conventions gave them the unprecedented opportunity to travel without their husbands. Unaccompanied, Gustie and Flora attended meetings not only in New York City, but also in Cincinnati, San Antonio, and Los Angeles. Following the Los Angeles convention in 1902, Flora Stanley embarked with several women friends on an extensive tour of the west, visiting Northern California, Oregon, Washington, Yellowstone, and the Grand Canyon. Her trip consumed her diary from April 19 to July 17, 1902 (Stanley Museum Archives).

While club society may have had greater impact on their lives, it was by no means the sole province of middle-class women. The men of Newton retired to their own single sex clubs when the workday was over. The Monday Evening Club and the Tuesday Club in Newton Corner were favorites of the Stanley brothers. It was before the membership of these organizations that Freel first presented his history of the Stanley dry plate business. He also lectured on the violin, the misguided mission of labor unions, and amused his fellows with tales of the Stanley twins' youthful enterprises.

The main difference between the men's and women's clubs came down to subject matter. Men favored technological, political, and philosophical topics, while women preferred to tackle social issues and master the arts. While the ladies debated female suffrage, the gentlemen tended to hunker down over US tariff policy. The local Tariff Reform Club, first organized in support of President Cleveland's plan for retrenching the nation's highly protective tariff system, drew the interest if not the membership of Frank Stanley. Though a faithful Republican on most ballot issues, F.E. supported the Democratic position on the tariff. The idea of protecting American industry through high tariffs on foreign competitors offended his Yankee belief in standing on ones own and beating the field with a better mousetrap. In a letter to Gustie dated July 25,1897, F.E. wrote that protective tariffs "don't protect anybody but a few monopolists who are rich and influential enough to shape legislation to suit themselves" (Stanley Museum Archives). Stanley's persistent anti-tariff sentiments even led him to vote once for Woodrow Wilson.

According to Gustie, Frank had a great flare for presenting papers. When he spoke as a guest lecturer on "airships" at her Social Science Club on April 20, 1910, she wrote in her diary that he had made a big hit: "He wore his new light suit, with lavender tie and shirt. He must have impressed everyone with the idea that he knew his subject and could answer anything they might ask...He had his little models to explain the different kinds of aircraft." Stanley, the serial technology buff, now had his curiosity piqued by the mechanics of flight. In 1907 he was one of the original incorporators of the Aero Club of New England, whose membership included fellow automobilists Charles Glidden and Henry Howard.

On an earlier speaking occasion, with his playful wit on display before the membership of the Katahdin Club, Frank had failed to tickle his wife. As the featured after-dinner speaker, he delivered a humorous history of the club which he claimed Gustie had written, and he had memorized. Her diary for February 11, 1907, concluded: "There were many funny things in it and some that were not so funny." Usually, the thorough diarist, on this occasion Gustie recorded no examples.

Thanks to Augusta, however, who privately published her husband's club papers in a posthumous volume called *Theories Worth Having*, we have numerous samples of Frank's weightier writings. From the contents of these densely packed essays, it is clear to see how profoundly Stanley was a man of the new technological age. Asserting that organized religion has served largely to hold back the progress of mankind, he argued for the need to supplant the teleological expectations of faith with the indeterminate process of science. He declared that any theory of man's nature or his origins must conform to the demonstrable facts and principles of science not "with the so-called revelations of ancient times" (F.E. Stanley 1919, 17).

Stanley's writings suggest that he probably read Thorstein Veblen. In an argument similar to Veblen's, Stanley identified scarcity as the source of struggle in the world and industrialization as the key to mitigating it. But the strongest influence on Stanley's thinking was Social Darwinism— the Gilded Age philosophy that touted survival of the fittest. While his language sounds clearly derivative of the social philosophers of his day, Stanley spun these ideas with a downeast accent.

Frank Stanley's interpretation of jungle law was leavened with a strong dose of Yankee work ethic: "...The organic growth which we call progress is the outcome of a struggle for existence where the fittest survive, and by the fittest is meant the most efficient in the particular struggle in which they are engaged" (F.E. Stanley 1919, 17). He added, that in competition between two merchants, with all other factors equal, "the honest man will win." The voices of several generations of civic-minded Stanley forebears can be heard in Frank's idiosyncratic version of industrial competition. In the Stanley lexicon competition is a highly evolved form of cooperation: "...(M)an finally discovered that when he met another man, instead of killing or eating him, he was more valuable alive, as an assistant..." (F.E. Stanley 1919, 169). In another essay entitled "The Science of Making a Living" Stanley continued this train of thought: "...(E)verybody as a general rule who is making a living is working for someone else. We supply our wants by supplying the wants of others." Astonishingly, Stanley managed to temper Social Darwinism with the Golden Rule by way of the New England town meeting.

When not theoretically engaged, Frank enjoyed the lighter side of club life. He liked playing a round of golf at the Brae Burn Country

Club in West Newton, while Brother Freel preferred relaxing with the Newton Camera and Bicycle Clubs. Freel was also the driving force behind the formation of the Newton Corner Hunnewell Club. Founded in 1895 on the bedrock of temperance, the Hunnewell was to be a place where friends and neighbors gathered to recreate and read. The club was similar in mission to the Newton Club in Newtonville, the Auburndale Club, and the West Newton Neighborhood Club. Membership in the club was originally limited to forty men living within a two-square mile area in Newton Corner.

At F.O.'s behest, twelve of these members met in November 1897 to incorporate the club. Stanley also enlisted his brother Frank to join him in purchasing a large lot on the corner of Church and Eldredge Streets. The brothers then built a three-story classical revival building with steel frame construction. Greek columns and a porch balustrade distinguished the façade, while hand-carved mahogany and gas lighting with brass Towle fixtures graced the interior. The original design included outdoor tennis courts, a first-floor billiards room, library, and reading room. There was a formal dining room with servants' kitchen located on the second floor along with card rooms and a parlor for the ladies, who could visit the club only on certain days. The third floor had four bowling alleys and an assembly hall that could seat several hundred people. An air of elegance prevailed at all times at the Hunnewell with doormen, stewards, and even pin setters in uniform. F.O. established a tradition—one that lasted into the 1930s—of Sunday afternoon concerts with tea at intermission (Stone 1972).

The membership held their first meeting in the new quarters in April 1898. At that time the Stanleys gave custody of the property to the members on a free lease rental basis (Stone 1972). In 1909, a new hall was added to the existing structure and named appropriately the "Stanley." It opened on February 22 with a formal ball—duly recorded in Gustie's diary. The four Stanleys led the grand march, with the "boys" (as Gustie called them) in strict evening dress and their wives each in white lace over yellow.

In 1912, Frank was elected president of the club, and Gustie thought it would be good for his "sociability." She confided to her diary that he has "staid (sic) at home too much." While she was pleased to get Frank into circulation, Gustie disdained the Hunnewell Club as a social institution. She wrote in her diary entry for March 6, 1911: "Whatever fine can be said about the H.C., it is not a social centre (sic). But all the new people go and on that account it helps to make the new residents more contented, and they have lovely times among themselves doubtless." Frank Dingley was right. the Stanleys had traveled a long distance "from the chimney corner in Auburn."

In all the years they lived in Newton, neither Francis nor Freelan Stanley ran for public office. They did not head up commissions or chair select committees. In fact, it might be said that their wives carried on the tradition of community service established by earlier generations of Stanleys. Perhaps Newton was just too far along its path to modernity to engage the brothers' interest. By contrast, each of them took a founder's pride in their vacation homes. Frank saw to the laying of sidewalks on Squirrel Island and was instrumental in designing and building the island's water storage system. But Freel, who went west for his health, is the one who most clearly embraced the ancestral imperative for civic participation.

Though smaller in scale, the Stanley Hotel in Estes Park recalls the grace and grandeur of the White Mountain resort hotels.

Stanley Museum Archives.

A view of the main lobby at the Stanley Hotel.

Stanley Museum Archives.

CHAPTER 17

The Better Mousetrap

FOR A NUMBER OF YEARS, F.O. Stanley's ongoing struggle with tuberculosis had ensured that he and Flora spent less time in Newton than did Frank and Gustie. While the latter were rising in the ranks of society on the East Coast, Flora and Freel became increasingly important to the life of their adopted western home in Estes Park. Although it might be assumed that a man in chronically ill health would do little more than enjoy the view, from the beginning of his time in Colorado, F.O. Stanley energetically engaged in town affairs. Like his grandfather and father before him, Stanley left his mark on the civic landscape. His achievements in Estes Park included the refurbishing of the town water supply, the building of a bank, a mountain road, and a hydroelectric plant. In Colorado, far from the shadow of his twin brother, F.O. Stanley helped bring a rustic western town into the twentieth century by transmuting its dwindling mining economy into tourist gold.

F.O.'s Colorado projects, however, were not undertaken on a strictly disinterested basis. He saw in his new surroundings the opportunity to build a hotel—rivaling such eastern beauties as the Mount Washington—that would serve as the nucleus for a mountain resort in the west. He knew that the success of this venture depended on whether Estes Park could provide a dependable source of electrical power, a good water supply, and a modern roadway to bring tourists from the railway station sixteen miles away. Barely one year after he first arrived to regain his health in the mountain air, the old Yankee got down to the business of road improvement.

In 1906 when the Colorado Good Roads Movement convened for the first time, it named F.O. Stanley vice president of the organization. Without wasting time, Stanley pledged himself to the task of getting a good road built from Lyons (where the Burlington Railroad had a terminal) to Estes Park. Since Colorado had no state department of roads at this time, Stanley had to generate the funds for the project. To that end, he contributed $5000 of his own money, wrested nearly the same amount from the Colorado legislature, and raised the rest by private subscription (Pickering 2000, 102). It was necessary to lay some new roadbed, but the route to Lyons hooked into the existing North St. Vrain Road, which was upgraded and rebuilt. Stanley's business hand is evident in the project. At intervals along the route he had concrete water reservoirs constructed to accommodate the operation of steam vehicles—especially those without condensers!

The roadwork was finished in the fall of 1907, and the county agreed to take it over. Shortly afterwards, construction began on Stanley's hotel. F.O. himself served as the seminal architect for the project which was drafted in the Stanley brothers' favorite colonial revival style. He hired Denver architect Robert Wieger to work on the plans, so it is impossible to tell how much of the final design was Freel's alone (Pickering 2000, 127).

During construction of the hotel, it became obvious to Stanley that the local water supply was insufficient to serve the needs of his hotel no less a growing resort town. Thus, while the building progressed, he joined with four other men in September 1908 to form the Estes Park Water Company—a public utility that would oversee both the installation of a new water main and the building of a reservoir (Pickering 2000, 151). Stanley did not stop there. Within a month, he was at work incorporating the Estes Park Light and Power Company which was responsible for building the town's first hydroelectric plant.

The Stanley Hotel, completed at a cost of nearly a half million dollars, opened for business in June of 1909. A fleet of twenty-two

Chansonetta Emmons took this photograph of F.O. Stanley's grand last Newton residence built in 1913. Photograph by Chansonetta Stanley Emmons.

Stanley Museum Archives.

Mountain Wagon Steamers (forerunners of the airport shuttle bus) met guests at the train station and motored them to the hotel. F.O. himself designed the mountain wagon—a nine-passenger vehicle using the thirty-horsepower engine developed for the Stanley racer. This new wagon once again demonstrated the steam car's grade climbing ability by carrying a large payload of passengers and baggage over the mountains. In fact the vehicle did the job so well that other resort owners in the Rockies began to make inquiries. To accommodate this potential new market, Freel opened a dealership for the SMCC at 1523 Cheyenne Place in Denver. According to Augusta Stanley's diary, F.O. left on April 12, 1910, "to start an agency" and "look after his hotels at Estes Park."

Judging by how briefly they visited, the Frank Stanleys did not share Freel and Flora's enthusiasm for the grandeur of the West. Indeed, Gustie and Frank preferred the decidedly less rustic environs of Squirrel Island where Frank piloted his boat *Empress* around the waters of Boothbay Harbor, and Gustie, increasingly plagued by knee and leg pain, enjoyed the rounds of bridge and summer gossip. They visited Estes Park together just once in August 1917. According to Gustie's lackluster journal accounts, the trip was hardly memorable, although it is difficult to distinguish Augusta's true impressions from her feelings for Flora. Gustie complained that the cuisine at the Stanley was "not up to an Eastern (sic) hotel" but allowed that their room was fine. On August 27, however, she assumed the note of competitive condescension she often adopted when discussing her sister-in-law: "Freel's folks seem to want us to be comfortable—but it is not in Flora's nature to be really hospitable." With a parting salvo in her journal she dismissed Rocky Mountain culture as "nothing ... very artistic" and hurried back to the more tasteful precincts of Squirrel Island. She revisited Estes Park on her own five years later, however, and did concede that the drive to the top of the Continental Divide was "beyond description."

Gustie and Frank "roughed it" at their Squirrel Island cottage. The Stanley twins had decidedly different taste in vacation retreats.

Stanley Museum Archives.

Freel and Flora continued to vacation in the Rockies every summer even after doctors gave F.O. clearance to resume his life in Newton. In 1907, the couple wintered in New England for the first time since 1903. Their return to a normal routine was later marred by a deadly accident. While driving in Malden, Massachusetts, on February 17, 1911, F.O. struck seven-year-old Vivian Black as she darted in front of his car. Freel himself drove the child to the hospital, where she died the next day. According to the *Newton Graphic* Stanley was arrested and arraigned on the charge of manslaughter. Gustie noted in her diary that Frank posted $1000 bail and later accompanied Freel and his lawyer Samuel Leland Powers to Malden District Court. F.O. testified that he had not exceeded seven miles per hour on the street. He saw the little girl cross safely in front of his car, but then she dashed back unexpectedly. He explained that as he tried to brake on the ice, the car skidded sidewise, causing the rear mudguard to strike the child. An eyewitness corroborated Stanley's account in every detail, and on March 24, the judge found him not responsible. He did fine Stanley $21.00 for driving with a lapsed license. On account of his lengthy stays in Colorado, F.O. was apparently unaware that a new form was necessary.

In happier days, Freel and Flora marked his return to good health by building a grand estate at 337 Waverly Avenue (now 50 Green Park). Freel lent a considerable hand to the design of the new house, which was conceived in the colonial revival style and built in brick. The final plans, drafted by architect James Purdon, included five symmetrical bays and one raised bay off the living room for musicales. There were two second-floor dormers, two rear chimneys, and a slate hip roof. Freel equipped the garage, as before, with a wooden turntable for his autos. The couple moved into their new quarters in September 1913.

The first time that Freel and Flora built an elegant home in Newton, Gustie and Frank followed suit, but this time Gustie had to content herself with remodeling her cottage on Squirrel Island. She had a hard time concealing her envy in the pages of her diary. On September 15, 1913, she wrote: "I do not like the brick work. It is the kind I never

Here seen blending with the surroundings F.O. Stanley, while recuperating in the Rockies, carved out a life quite independent of his brother.

Stanley Museum Archives.

F.O. Stanley lived twenty-two years longer than his twin. Though he never again embarked on a commercial mechanical project, he found solace making violins with his nephew Carlton. Photograph by Chansonetta Stanley Emmons.

Stanley Museum Archives.

liked." June 13, 1914: "Went through Freel's house today. I wouldn't swap with them by a good deal." May 2, 1915: "They have put great expense into it and it is very pretty. But oh! So much care!" One suspects that Gustie would have been delighted to take on the trials of a house bigger than Flora's if she and Frank had had the funds. But by 1913 the Stanley Motor Carriage Company had been struggling financially for several years. As the partner on the scene, Frank appears to have borne the brunt of responsibility for the declining business. By contrast, Freel probably stood on the soundest financial footing of his life due to his Estes Park enterprises.

Although the twins had never realized the kind of profits from the sale of steam cars that they had enjoyed in the dry plate business, they expected at the very least to break even. There was cause for concern, therefore, when the company nearly ran into the red in 1910. Interestingly, the year started out with the promise of good sales. As Freel left Newton to open a Denver franchise in the spring of that year, Gustie noted in her diary that "(t)he business seems to go on and the cars are being got out and into the market fast." Just three months later, however, the news she reported was very different. During July holidays on Squirrel Island, she twice referred to "business meetings" that Frank had called with his sons-in-law. On the thirty-first, she noted that he had sharply criticized Ed and Prescott for not "giving their undivided attention to the business." In the same entry she wrote, "the business is in straits and Frank expressed his ideas...I am afraid even that will not produce profits this year."

The next year brought no better outlook for sales. On January 12, 1911, Gustie confided to her diary that Frank and Prescott had had a serious fight at the factory and Prescott stormed out, staying away from work for two days. Presumably, he and Frank had argued over the faltering business. By April, prospects for the year had taken a turn for the worse. On the ninth of the month Gustie wrote, "I am feeling badly about the business, as I find Frank has put up a great many of his bonds to enable them to get through the winter." Just how much money Frank infused into the business is not known.

Newton Fire Chief W.B. Randlett sits next to his son, C.B. Randlett (at wheel) in a 1913 Model 78 Stanley with Fire Department insignia, extinguisher, and bell.   *Photo courtesy of Allen & Janette Blazick.*

Sales figures for the SMCC continued to be depressed over the next five years. At the Boston auto show in March 1913, Gustie sat in one of the Stanley exhibition cars to watch the crowds: "It makes me feel badly (sic) to see the people—the nice looking ones go right past our cars without deigning to look at them." She concluded that the Stanley models were "not stylish enough." In truth, Gustie's lament came at least a decade late. The steam automobile had been out of favor with the general buying public for more than ten years. Before 1900, at least 124 steam car companies were organized in the United States. Within two years the overwhelming majority of them had folded. By 1905, the SMCC and the White Sewing Machine Company of Ohio accounted for ninety per cent of steam car sales in America. White stopped building steamers in 1910, but Stanley sales figures did not improve as a result.

Cadillac probably dealt the knockout blow to the steam car market in 1912 by building a gasoline model that did not have to be cranked. Gas car manufacturers had been working for a number of years to perfect the self-starter, but Cadillac was the first to deliver one as a reliable standard feature. Once the motoring public could regularly start a car without cranking, it was hard to convince them that steam offered any appealing alternative. So great was the impact of the self-starter that when Abner Doble began building steamers in Detroit in 1913, he worked hard to make a self-starting model. Doble still believed in the superiority of steam automobiles. But his models while mechanically well built, cost a lot of money and took a minute and a half to start up. Even the Stanleys, however belatedly, attempted to counter the impact of the gasoline self-starter. In its 1916 company brochure, the SMCC recommended that motorists keep their pilot lights burning over night, "so that the car is ready at a moment's notice." The pilot could keep the steam up and hold for three to four days at a time. But persuading motorists, already apprehensive about the safety of boilers, to keep a pilot light burning all night in the garage proved a hard sell indeed (Bird and Montague 1971, 163-4).

The internal combustion engine alone, however, was not responsible for the Stanleys' business woes. In truth, the SMCC had begun to founder even before Cadillac's coup. The brothers' own miscalculation of the steam niche market—their singular unwillingness to accommodate the small number of consumers who still preferred steam cars—contributed to their slumping sales. The Stanleys' stubborn refusal to switch to kerosene burners (not until 1912) or to build condensers into their models (the first in 1915) doubtlessly cost them business. The hardheaded quirkiness for which the twins were justifiably known clearly worked against them in this instance. By refusing to concede that the customer's peace of mind and convenience were part of building a better product, the brothers eventually alienated their own clientele. When the Stanleys finally got around to building steamers with condensers, flaws in the design impaired the performance of the automobiles. Unless the condenser and water tank were frequently cleaned, heavy cylinder oil tended to foul the boiler. Thus after many years of a "well-merited reputation for service at low cost," Stanley dependability was compromised (Doble, 1916). Furthermore, the redesign of the Stanley line to include condensers effectively cut production in 1915 to a mere 126 automobiles (R. Stanley 1943, 17).

It might fairly be stated that the British War Office, which needed steam power to fight the Great War, kept the SMCC afloat during 1915. Even before America entered the fray, the SMCC contracted with England to produce compact boilers and burners for use at the front lines. Stanley units pumped water from the trenches and provided mobile showers for men who were victims of mustard gas attacks.

Despite the infusion of cash from overseas contracts, the war years were tense and unhappy ones in the Frank Stanley household. While F.E. struggled to save his failing business, Gustie fretted that their only son Raymond would have to serve abroad. In the meanwhile she joined other American women volunteers in sewing bandages for the Red Cross and knitting woolen socks for soldiers at the front. Color blindness kept Raymond out of the Air Corps and safely at a government job in Washington, D.C., for the duration. But matters did not turn out as well for Frank. The beginning of the end of Stanley control of the SMCC came in 1916. Augusta recorded in her diary on September 14: "Frank is very much upset over business. It appears that they have been losing money all summer—in fact the past year, and he thinks they will have to go out of business or lose all they have." The following month on October 27, she noted that "the men are here to see about buying the automobile business." By February 2, 1917, Gustie and Frank hosted an engineer whose job it was to report the state of the company to potential investors, a Chicago-based consortium headed by Charles Counselman and Arthur Goodwillie. The deal closed on March 25, 1917, and Gustie marked the day: Frank is really "out of business," she wrote.

That spring, the Stanley Brothers settled accounts with each other as they had always planned to do. Since neither twin had ever received a salary from the SMCC but instead had withdrawn funds from the business when needed, the "grand settlement," entailed dividing the remaining assets as fairly as possible. Although the actual amount of

F.E. Stanley. Notman Studios Photo.
Stanley Museum Archives.

each brother's share is not known, Emma Walker, the long-time office manager for the SMCC, wrote Raymond Stanley in a letter dated February 7, 1937, that F.E. took out several thousand dollars more than F.O. over the lifetime of the business (Stanley Museum Archives). According to Walker, Frank used more company funds to finance the various homes he built for his daughters than F.O. did to build his hotel and other projects in Estes Park. Walker stated that in the settlement F.O. accepted Massachusetts State bonds from his brother in lieu of cash, although the bonds were less valuable. Somewhat puzzlingly, she made no reference to the personal funds Frank supposedly funneled into the SMCC to keep it going.

The Chicago consortium that purchased the nearly bankrupt Stanley Corporation refinanced the business in the amount of $2,500,000. Both twins withdrew from management, but Ed Hallett, Prescott Warren, and Carlton Stanley were retained to run the company—at least for a time. Several months into the new management, however, changes were in the offing. Gustie reported in her diary on May 30 that her sister Emma Walker had resigned her position at the factory under duress. Prescott Warren was eventually replaced as company president. Ed Hallett continued to serve as corporate treasurer and Carleton Stanley—the invaluable nephew—stayed on as vice president. F.O. paid no further official attention to company business, but F.E. allegedly served in an advisory capacity. Even that diminished role vanished by Thanksgiving time. Gustie recorded in her diary the resultant strain on family relations at the holiday dinner table: "We feel so badly about things the boys (Prescott and Ed) have done toward Frank that there seems no pleasure for any of us. But we tried not to show it."

The new management of the SMCC had no luck reigniting steam car sales. Production figures never broke the low hundreds, forcing the Chicago group into receivership by 1923. An organization called the Steam Vehicle Corporation of America bought out the original consortium. Three years after that, a concern known as Stanley Steamer, Inc. purchased the company's assets, patents, drawings, and records in

F.O. Stanley. Notman Studios Photo.
Stanley Museum Archives.

a futile effort to revive the business. In 1927, the last sixteen Stanley steam vehicles were built in the United States. The American steam car industry as a whole quietly expired five years later when Abner Doble quit making cars in 1932.

At the time Frank Stanley personally withdrew from the steam car business, Gustie issued wifely concern that he might not have enough to do "to feel comfortable." She need not have worried. Even before the demise of the SMCC, he and Freel (with young Raymond having a modest share in the business) had partnered again, this time contracting with the US government to develop inter-urban transportation (Crank 1995, 12). The type of vehicle they planned to build was a steam-powered unit rail car, economical for spur lines and short runs. The chief engineer-designer on the project, a man named P. H. Gentzel, lived in Freel's Jefferson Street house so that he could work closely with the Stanleys while designing the rail car. By fall of that year, the Stanley Unit Rail Company, which built its cars in Concord, New Hampshire, had one ready to test. On October 11, 1916, Gustie and Frank took a twenty-mile ride on the Boston and Maine line from Bethel to Rochester, Vermont. Gustie gave the rail car a positive review, writing in her diary, "I wouldn't have known that I was not on a regular locomotive train it ran so nicely." Her assessment appears to have been a minority opinion. The four cars produced by the Stanleys were variously criticised as "hard riding and unreliable " (Haartz 1996, 25).

On July 31, 1918, according to his own handwritten itinerary, Frank Stanley set out from Boothbay Harbor for White River Junction, Vermont, presumably on business for the unit car project. For whatever reason, he changed his mind en route. After reaching Portland, he did not turn west but continued to travel south, perhaps intending to stop at his daughter Blanche's summer home in Wilton, New Hampshire. But he never did. Instead, Stanley was traveling towards Boston on the Newburyport Turnpike at his usual speedy clip.

Years earlier, he had said of his fast driving that "when riding at the highest speed at which I ever run on the road, it never occurs to me that there is any danger, not even that the other fellow will get there first." On this summer day, however, sixty-nine-year-old Frank Stanley found more trouble than he even could outrun. As he came over the crest of a long hill, he had to swerve to avoid colliding with some vehicles stopped on the roadway. Unable to regain control, he first ran off the pavement, then veered back to the southbound side of the road, traveled up an embankment and flipped the car over, thus pinning himself underneath the steering wheel. Stanley died of a skull fracture in transport to the Beverly Hospital. The medical examiner told Mrs. Stanley that there were no broken bones or even marks on the body. Apart from a crushed windshield, F.E.'s custom built, five-passenger-steamer, likewise showed no sign of the seriousness of the crash. Frank's body was brought back to his home in Newton and the casket placed in the reception room where Blanche and Ed Hallett had received their wedding guests years before (B. Hallett 1954).

Newspaper notices of Stanley's death could not resist marking the grim irony of his fate. "HIS OWN AUTO KILLS INVENTOR" ran the headline of the Newton *Town Crier*. Mayor Childs of Newton delivered a more-measured response to the widow in a condolence letter. He hailed F.E. as one of the leading citizens of the city: "…(I)t has always seemed to me that his genius, his industry, his capacity for work and his honesty really did lead younger men to sense the opportunity which opened before them…He certainly showed the world the way to do things…His democratic ways in the midst of his successes proved him a true American…(Stanley Museum Archives).

Frank left Augusta (according to her diary notation on October 30, 1918) an estate worth about $400,000—a sizeable amount for the day—plus stock of dubious value in the Stanley Motor Carriage Company and the Unit Rail Car Company. By comparison to her former style, however, Gustie was forced to live more modestly on $1500 a month. Long accustomed to affluence, she found herself having to make many little economies that seemed daunting to her. She now took the train into Boston and rode the electric cars in Newton. She had to share the services of her chauffeur Wallace with Freel and Flora. Given the difficulties Gustie had walking in her later years, the loss of ready transportation no doubt proved a hardship for her.

The Stanley daughters also struggled to maintain the lifestyles they had enjoyed before their father's death. When Ed Hallett lost his job at the SMCC in 1924, this put a serious financial strain on Blanche's household. Two years later, Prescott Warren died and Emmie had to sell her house and move in with her mother. When two of Gustie's granddaughters had their coming-out parties together in 1927, the budgetary pinch was palpable. Blanche Stanley Hallett wrote, "I didn't have enough money to afford all the dresses Augusta (her daughter) and I had to have. One store in Boston made it very unpleasant for me. I can't help being amused when I remember that my mother had always told me that one day I would be a rich woman" (B. Hallett 1954). Augusta Stanley lived nine years after Frank's death and died at her Centre Street home in 1927.

At the time of Frank's death, Freel told Gustie that he would never recover from the loss of his twin, and in some ways that appears to have been so. The collapse of the steam car business had boded ill for the Stanleys' productive partnership, but the extraordinary collaboration between the twin brothers ended for all time with Frank's fatal accident. Freel survived twenty-two years without the company of his twin but never ventured again into any mechanical enterprise. To all outward appearances, however, he and Flora led lives of contented retirement. Perhaps spurred on by his brother's sudden death, F.O. continued to liquidate his real estate holdings in Estes Park, though the couple returned for many summers to the mountains (Pickering 2000, 232-233). For a man long plagued by tuberculosis, F.O. Stanley enjoyed remarkably robust heath in old age, surviving to age ninety-one. Perhaps the secret of his vigor lay in the pleasure he derived from making violins in his Waverly Street garage with his nephew Carlton.

On October 2, 1940, a year after Flora had died of a stroke, F.O. walked outside for his morning newspaper, collapsed and died shortly thereafter.

Before his death, F.O. Stanley left some final thoughts on the technological battle between steam and internal combustion. Though he betrayed not the slightest regret for having spent a lifetime building steam cars, he concluded in a letter dated April 27, 1939, that now "all the time and all the money spent in trying to make a steam car as good, or better, than a gas car, is time and money wasted." Self-described as the man who "has had more experience in making steam cars than any other man living in the world today," he conceded that the internal combustion engine, which he called "one of the most valuable inventions made by man," had proved to be the better mousetrap (Stanley Museum Archives).

American industry does not celebrate its losers. Those inventors who guess wrong at the start of a new technological age usually wind up in history's dustbin—no matter how ingenious the technology they practice. Why, then, does anyone remember or should anyone remember the Stanleys, self-admitted also-rans in the automobile industry?

Even in their own time, an aura of nostalgia enveloped the twin brothers. To their fellow Americans rushing pell-mell into a different arrangement of society and industry that "great pair of Yankees," as Charles Abbott called them, already represented a link with the rapidly receding past. Amidst the pioneers of the mass market—the Eastmans, Fords, and other industrial giants of the Gilded Age—the Stanley Brothers did business like New England tradesmen. Hewing to the simple code espoused by Benjamin Franklin that held a man needs nothing more than public trust and a superior product to sell his wares, they managed for a time to buck the emerging trends with a high degree of personal success. During an era of rapacious capitalism the Stanleys practiced moderation. In a time of ruthless individualism, they worked to build community. The example of these Maine brothers seemed to offer hope that the new age of business might yet be rooted in the virtues of the early republic.

One hundred years later at the outset of the early twenty-first century, when a worldwide resurgence in unfettered capitalism has rendered the business ethic of Franklin ever more remote, the Stanley Brothers still strike a cord with their fellow Americans. Perhaps the startling similarities between the Gilded Age and our own times have stirred doubts and concerns that make the Stanleys' tale of sturdy self-reliance again inspirational. For it is fair to say that the computer and the microchip have unsettled our notions of time and space as profoundly as the telephone, the electric light, and the automobile did more than a century ago. With the dawning of the computer age—just as in the first days of the automotive industry—has come a familiar welter of competing technologies and entrepreneurial activity, with the same attendant spate of failed bids for domination in the field.

What seems clear is that in any technological revolution once the possibility is known, the battle ensues over format. The story of the Stanleys' brief business success—for all the quirky singularity of the brothers themselves—exemplifies the kind of redundancy that results when one format supercedes another. The steam mechanics the brothers practiced was sound science, but they lost out to internal combustion through a combination not only of their own business decisions and but also of events entirely beyond their control. In our own time, the transcendence of VHS technology over the Beta Max system and the triumph of Microsoft's operating system over all comers evoke the Stanleys' losing war with the gasoline engine.

REFERENCE LIST

Bacon, John H. 1984. *American Steam Car Pioneers*. Eaton, Pa.: The Newcomen Society of The United States.

Balasco, Warren James. 1979. *Americans on the Road*. Cambridge, Ma.: MIT Press.

Batchelder, A.G. 1906. Carnival on Florida beach. *Automobile Magazine* 14 (1 February): 271, 274, 278.

——. 1907. Florida meet spoiled by freak sprinter. *Automobile Magazine* 16 (31 January): 247-248.

Bird, Anthony and Lord Montagu of Beaulieu. 1971. *Steam Cars 1770-1970*. New York: St. Martins Press.

Brayer, Elizabeth. 1996. *George Eastman: A Biography*. Baltimore: The Johns Hopkins University Press.

Brimblecom, John C., ed., 1914. *Beautiful Newton*. Newton, Ma.: Newton Graphic Publishing Co.

Brown, Richard D. 1976. Modernization: a Victorian climax. In *Victorian America*, ed. D.W. Howe. Philadelphia: University of Pennsylvania Press.

Burt, F. Allen. 1960. *The Story of Mt. Washington*. Hanover, N. H.: Dartmouth Publications.

Collins, Herbert Ridgeway. 1971. *Presidents on Wheels*. Washington, D.C.: Acropolis Books.

Coons, Clarence F. 1991. Stanley steamers in Maine in the 1912-1914 era. *Stanley Museum Newsletter* 10(1): 9-19, 19-21.

Cooper, Jonathan. 1986. Violin making in the United States pt. 1. *Stanley Museum Newsletter* 5(3): 6.

Crank, James D. 1997. The steam car in America. *Stanley Museum Quarterly* 16(3): 14-17.

——. 1995. Data on the Stanley unit rail car. *Stanley Museum Quarterly* 14: 12.

Davis, Susan S. 1992. F.E. Stanley, drawing and the airbrush. *Stanley Museum Newsletter* 11(2): 12-13, 20.

——. 1992. Fred Marriott, king of speed and a tradition's link. *Stanley Museum Newsletter* 11(2): 3-7, 16.

——. 1997. *The Stanleys: Renaissance Yankees*. New York: The Newcomen Society of the United States.

Derr, Thomas S. 1932. *The Modern Steam Car and Its Background.* Boston: Blanchard Print Co.

Doble, Abner. 1916. Steam motor-vehicles. Paper presented to the Cleveland Section Meeting. 20 October. Reprinted at www.railroadextra.com/automo.html.

Eastman, George. 1994. Correspondence with F.O. and F.E. Stanley. George Eastman House Archives. Reprinted in *Stanley Museum Quarterly* 13(2): 12-13, 16-20.
——. Correspondence. George Eastman House Archives, Rochester, N.Y.

Elliott, Harmon. 1941. *The Story of a Father and Son.* Cambridge, Ma.: The Elliott Addressing Machine Co.
——. 1945. *The Sterling Elliott Family by His Only Son.* Cambridge, Ma.: The Elliott Addressing Machine Co.
——. [1941] 1994. An invention that will live forever. Reprinted in the *Stanley Museum Quarterly* 13(3&4): 14-16.

Emmons, Chansonetta Stanley. [1916] 1992. Stanley family. *Stanley Museum Newsletter* 11(4): 5-13.

Fleishman, Thelma. 1999. *Images of America Newton.* Charleston, S.C.: Arcadia Publishing.

Gardner-Huggett, Joanna. 1996. Chansonetta Stanley Emmons: Photography of Europe and the Carolinas. *Stanley Museum Quarterly* 15(3&4):7-9.

Gordon, Albert I. 1959. *Jews in Suburbia.* Boston: Beacon Press.

Haartz, Eric. 1996. The Stanleys' unit rail car in New England. *Stanley Museum Quarterly* 15(3&4): 24-25.

Hallett, Blanche Stanley. Memoirs. 1954? MSS. Stanley Museum Archives. Kingfield, Me.

Handlin, Oscar. 1959. *Boston's Immigrants.* Cambridge, Ma.: the Belknap Press of Harvard University.

Hart, Arthur C. 1985. Climbing a hill called Dead Horse. *Stanley Museum Newsletter* 4(2): 9.
——. 1992. Stanley vaporizers, fuel, and good steaming. *Stanley Museum Newsletter* 11(1): 10-12.

Hendrickson, Robert. 1979. *The Illustrated History of America's Great Department Stores.* New York: Stein and Day.

Hofstadter, Richard. 1971. *America at 1750, a Social Portrait.* New York: Knopf.

Howe, Daniel Walker. 1976. Victorian culture in America. In *Victorian America*, ed. D. W. Howe. Philadelphia: University of Pennsylvania Press.

Jenkins, Reece V. 1975. *Images and Enterprises, Technology and the American Photographic Industry 1839 to 1925*. Baltimore: The Johns Hopkins University Press.

Katz, John F. 1987. F.E. & F.O. Stanley; the challenge from steam. *Automobile Quarterly* 25(1): 14-29.

Mason, Philip P. 1957. *The League of American Wheelmen and the Good Roads Movement, 1880-1905*. Ph.D. diss. University of Michigan. Ann Arbor.

Marshall, Thomas C. [1982] 1996. Interview with the late Fred Marriott. Reprinted in *Stanley Museum Quarterly* 15(3&4): 20-23.

McKay, William W. 1986. Wisps of early steam pt. 1. *Stanley Museum Newsletter* 5(2): 8.
——. Wisps of early steam pt. 2. *Stanley Museum Newsletter* 5(3): 8-9.

Merrick, H. James. 2001. Archivist's report. In the *Stanley Museum Quarterly* 22(2): 6.

Morison, Samuel Eliot. c. 1962. *One Boy's Boston 1887-1901*. Cambridge: The Riverside Press.

Nevins, Allan. c. 1927. *The Emergence of Modern America, 1865-1878*. New York: The Macmillan Company.

Newton Department of Planning and Development and the Newton Historical Commission. 1978. *Newton's 19th Century Architecture: Newton Corner and Nonantum*.

Ochsner, Jeffrey Karl. 1988. Architecture for the Boston and Albany Railroad 1881-1894. Reprinted from *The Journal of the Society of Architectural Historians*. 47 (June): p. 109.

Parrington, George C. 1889. *History of the State Normal School, Farmington, Maine*. Farmington, Maine: Knowltoon, McLeary & Co.

Pearson, Norman H. 1992. The Chansonetta Yet to be Discovered. *Stanley Museum Newsletter* 9(1):6.

Pickering, James H. 2000. *Mr. Stanley of Estes Park*. Kingfield, Maine: The Stanley Museum of Maine.

Punnett, Dick and Yvonne. 1992. Thrills Chills and Spills 1906-1929. New Smyrna Beach, Florida: Luthers.

Punnett, Dick. 1997. *Racing on the Rim*. Ormond Beach, Fla.: Tomoka Press.

Rowe, Henry K. 1930. *Tercentenary History of Newton 1630-1930.* Cambridge, Ma.: Murray Printing Co.

Schlereth, Thomas J. 1991. *Victorian America: Transformations in Everyday Life. 1876-1915.* New York: Harper Collins Publishers.

Schlesinger, Arthur Meier. c. 1933. *The Rise of the City.* New York: The Macmillan Company.

Smith, Page. 1966. *As a City on a Hill.* New York: Knopf.

Stanley, Augusta. 1906-1927. Diaries. Stanley Museum Archives. Kingfield, Me.

Stanley, Flora. 1884,1886, 1888-1890, 1895-1896, 1889, 1901-1903, 1910. Diaries. Stanley Museum Archives. Kingfield, Me.
——. [1899] 1999. The first motor carriage ascent of Mt. Washington, August 31, 1899. In *Centennial of the first auto up Mt. Washington,* Susan S. Davis,11-16, Kingfield, Me.: Stanley Museum.

Stanley, Francis Edgar. [1914] 1991. US Senate 25 May 1914. Anti-trust hearings on Eastman Kodak. Reprinted in the *Stanley Museum Newsletter* 10(3): 12-13.
——. 1903. Correspondence with George Eastman. Eastman House Archives. Rochester, N.Y.
——. 1907. Letter to the editor. *Motor Age* 11(1) (3 January): 27.
——. 1994. Correspondence with George Eastman. George Eastman House Archives. Reprinted in the *Stanley Museum Quarterly* 13(2): 16-20.
——. 1919. *Theories Worth Having, and Other Papers.* Boston: Priv. print.
——.Correspondence. Stanley Museum Archives. Kingfield, Me.

Stanley, Freelan Oscar. [1930] 1991. Maple syrup. *Stanley Museum Newsletter* 10(1): 13-14.
——. [1936] 1987. The Stanley dry plate pt. 1.
Stanley Museum Newsletter 6(1): 9, 14, 16.
——. [1936] 1987. The Stanley dry plate pt. 2. *Stanley Museum Newsletter* 6(2): 8-9.
——. [1936] 1987. The Stanley dry plate pt. 3. *Stanley Museum Newsletter* 6(3): 8-10.
——. 1994. Correspondence with George Eastman. George Eastman House Archives. Reprinted in the *Stanley Museum Quarterly* 13(2): 12-13, 16-20.
——. Correspondence. Stanley Museum Archives. Kingfield, Me.

Stanley, Raymond W. 1943. Steam lore. *Bulb Horn* 4(1) (January): 17-18.
——. 1945. Steam lore. *Bulb Horn* 6(3)(July): 18-19.
——. 1960. Chansonetta Stanley Emmons. For *American Heritage Magazine.* Reprinted in the *Stanley Museum Newsletter* 9(1): 10-12.

———. 1963. Evaporating the Stanley Steamer myth. In *Automobile Quarterly* 2(2): 120-129.

———. 1983. Family history. Mss reprinted in the *Stanley Museum Newsletter* 2(2): 4.

———. Correspondence. Stanley Museum Archives. Kingfield, Me.

Stilgoe, John R. 1988. *Borderland Origins of the American Suburb 1820-1939*. New Haven: Yale University Press.

Stone, Mason H. Jr. 1972. History of the Hunnewell Club of Newton. Mss at Jackson Homestead. Newton, Ma.

Sweetser, M. E. 1889. *King's Handbook of Newton*. Boston: Moses King Corporation.

Trachtenberg, Alan. 1982. *The Incorporation of America: Culture and Society in the Gilded Age*. New York: Hill and Wang.

Villalon, L. J. Andrew. 1981. The birth of an early automobile company. *Bulb Horn* 42 (2) (April/June): 11-21.

———. 1986. Strategies for success and survival in a changing marketplace, part 1, the Locomobile Company of America. *Bulb Horn* 47 (3) (July/September): 16-28.

Watertown Historical Society. 1900. *Watertown Records* vol 2. Watertown, Ma.: Press of Fred G. Barker.

Walker, Emma. [24 May 1936]. Letter to Raymond Stanley. Reprinted in the *Stanley Museum Newsletter* 6(3): 10.

White, John R. 1991 Review of *A History of Maine Built Automobiles* by Richard A. and Nancy L. Fraser. In *The Boston Globe* November 23, 1991. Reprinted in the *Stanley Museum Newsletter* 10(4): 13-14.

Whitney, George Eli. 1956. Papers. Smithsonian Archives.Washington, D.C.

Wiebe, Robert H. 1967. *The Search for Order 1877-1920*. New York: Hill and Wang.

Connor, Kathy. 2 March 1999 and 8 April 1999. Correspondence with the author. Curator, George Eastman House. Rochester, N.Y.

Hague, Nora. 17 February 2000. E Correspondence with author. Notman Photographic Archives, McCord Museum of Canadian History. Montreal.

AUTHOR ANONYMOUS ARTICLES

American Machinist. 1898. 21 (48) (1 December): 896-26.
21 (50) (15 December): 31-941.
21 (52) (29 December): 19-969 to 23-973.

Among the Automobilists. [1904] 1991. *Lewiston Evening Journal.*
Reprinted in the *Stanley Museum Newsletter* 10 (2): 3-4, 15.

Automobile Magazine. 1903. 8 (7 February): 181.
8 (28 March): 363.
1906. 14 (25 January): 265.
14 (1 February): 271, 274, 278.
14 (14 June): 943.
14 (31 May): 857-858.
15 (6 September): 297.

Record Smashed. [1904]. 1994. *Automobile Weekly*
Reprinted in the *Stanley Museum Newsletter* 13(1): 11-14.

Horseless Age, The 1896. 2(2) (December): 5.
1897. 2(5) (March): 7.
2(6) (April): 13.
1898. 3(7) (October): 7, 44.
3(8) (November): 10, 14.
3(9) (December): 22.
1899. 2(11) (February): 18.
4(16) (19 July): 6, 12.
4(25): (20 September): 6.
5(11) (13 December): 7-8.
1900. 6(17) (25 July): 24.
1901. 8(18 September): 528.
1902. 9(5) (29 January): 130-131, 151.
1904. 13 (9): 284.
1908. 21 (24): 694.

Motor World, The 1904   (March 3): 1005.

Scientific American. 1863. 8(March 14): 165.
1899. 81 (14 October): 282.
1903. 88(January 3): 3.
1903. 88 (11 April): 286.
1903. 88 (31 January): 72.
1904. 90 (25 June): 498.
1906. 94 (3 February): 115
1907. 96 (9 February): 128.

Stanley Museum Quarterly. [1904] 1991. 10(2): 15-16.

ELECTRONIC SOURCES

Adcock, Sylvia. Age of the auto . Driving in the fast lane. *Long Island Our Story.* http://www.lihistory.com.

Biographical Directory of the United States Congress. http://bioguide.Congress.gov/scripts/biodisplay.pl?index=D00356.

Boston Society of Mechanics History. 2001. http://www.b-s-m.org/history.html.

DeLong, Tedd. Charles Jasper Glidden. http://www.vmcca.org/bh/cjg.html.

Engineering News-Record. 2/1/99. http://www.enr.com.

ftp://ftp.rootsweb.com/pub/usgenweb/me/androscoggin/newspapers/press/mepress1.txt. Courtesy David C. Young and the Androscoggin Historical Society.

Henry Ford and racing–into the future. http://media.ford.com/article_display.cfm?article_id=7242.

Historical timeline of concrete. UNL website. Institute of Agriculture and Natural Resources.

Library of Congress, American memory. http://memory/loc.gov/ammem/today/nov28.html.

Shepler, John. 2001. In the heat of invention. http://www.execpc.com/˜shepler/steam.html.

Tech Notes. http://hvacwebtech.com.

Wright, Richard A. c. 1996-2001. http://www.theautochannel.com/content/mania/industry/history/chap 2.html.

MICROFILM

Postmasters, 1832-September 30, 1971 (Washington, D.C.: 1997).

Post Office Department Reports of Sited Locations 1837-1950 (Washington, D.C.: 1986).

INDEX

Stanley Museum Publications

1. *The Stanley Family Reunion: a transcript of conversations during the Stanley family gathering, June 7, 1981 at Kingfield, Maine* edited by Dan and Susan S. Davis (Kingfield, Me.: Stanley Museum, © 1982).

2. *Reflections on Transportation and Communication: An Evening with R. Buckminster Fuller* edited by Susan S. Davis (Kingfield, Me.: Stanley Museum, © 1983).

3. *Historic Touring: Early Tales of Steam Travel* edited by Susan S. Davis (Kingfield, Me.: Stanley Museum, © 1984).

4. *The Stanley Family* by Chansonetta Stanley Emmons (Kingfield, Me.: Stanley Museum, © 1993). *(Reprint of Chansonetta's 1916 genealogical article prepared for the Town of Kingfield's centenary.)*

5. *The Genealogy of the "Locomobile" Steam Carriage, 1899-1904* by Donald L. Ball (Kingfield, Me.: Stanley Museum, © 1994).

6. *The Stanleys: Renaissance Yankees, Innovation in Industry and the Arts* by Susan S. Davis (Exton, Pa.: The Newcomen Society of the United States, © 1997).

7. *A History and Tour of the Stanley Hotel, Estes Park, Colorado* by Susan S. Davis (Kingfield, Me.: Stanley Museum, © 1999).

8. *Commemorative Program – Centennial of the First Auto up Mount Washington* by Susan S. Davis (Kingfield, Me.: Stanley Museum, © 1999).

9. *Mr. Stanley of Estes Park* by James H. Pickering (Kingfield, Me.: Stanley Museum, © 2000).

10. *The Stanley Steamer: America's Legendary Steam Car* by Kit Foster (Kingfield, Me.: Stanley Museum, © 2004).

11. *"We Will Try This Hill" - The Climb to the Clouds 1904-1905* by H. James Merrick (Kingfield, Me.: Stanley Museum, © 2004).

12. *Bravo, Stanley!: The Racing History of Stanley and the 1906 Stanley Land Speed Record* by H. James Merrick (Kingfield, Me.: Stanley Museum, © 2006).

13. *The Old Table Chair*, by Chansonetta Stanley Emmons, (Kingfield, Me.: Stanley Museum, © 2009) *(Reprint of Chansonetta's 1909 booklet.)*

14. *The Stanley Museum Newsletter*, 1981-1992, superseded by *The Stanley Museum Quarterly*, 1993-present.

15. *the Winker*, the Stanley Museum's eNewsletter, 2008-present.

16. *Documenting a Myth: the South as Seen by Three Women Photographers, Chansonetta Stanley Emmons, Doris Ulman, Bayard Wooton, 1910-1940* a museum exhibition curated by Naomi Rosenblum and Susan Fillen-Yeh (Portland, Ore.: Reed College, © 1998).

You may purchase these publications electronically, via the Stanley Museum's website, www.stanleymuseum.org, by directing your browser to the "Museum Shop" link.

The Stanley Museum is pleased to offer its members and friends the opportunity to endow the Museum Archivist and the Executive Director positions. Please contact the Stanley Museum Director for additional information.

www.ingramcontent.com/pod-product-compliance
Lightning Source LLC
Chambersburg PA
CBHW060306100426
42742CB00011B/1885